SOME RECENT ADVANCES IN
MATHEMATICS
AND STATISTICS

SOME RECENT ADVANCES IN
MATHEMATICS
AND STATISTICS

Proceedings of Statistics 2011 Canada/IMST 2011-FIM XX
Montreal, Canada,   1 – 4 July 2011

Editor

Yogendra P Chaubey
Concordia University, Canada

 **World Scientific**

NEW JERSEY • LONDON • SINGAPORE • BEIJING • SHANGHAI • HONG KONG • TAIPEI • CHENNAI

Published by

World Scientific Publishing Co. Pte. Ltd.

5 Toh Tuck Link, Singapore 596224

USA office: 27 Warren Street, Suite 401-402, Hackensack, NJ 07601

UK office: 57 Shelton Street, Covent Garden, London WC2H 9HE

British Library Cataloguing-in-Publication Data
A catalogue record for this book is available from the British Library.

SOME RECENT ADVANCES IN MATHEMATICS AND STATISTICS
Proceedings of Statistics 2011 Canada/IMST 2011-FIM XX

ISBN 978-981-4417-97-6

Printed in Singapore by World Scientific Printers.

PREFACE

The present volume consists of 19 papers dealing with current issues in mathematical and statistical methods that were presented at Statistics 2011 Canada: 5th Canadian Conference in Applied Statistics that was held together with the 20th conference of the Forum for Interdisciplinary Mathematics titled, "Interdisciplinary Mathematical & Statistical Techniques" from July 1 to July 4, 2011. It was jointly hosted by the Department of Mathematics and Statistics, the Department of Decision Sciences & MIS of Concordia University, and the Forum for Interdisciplinary Mathematics (FIM) - an India based international society of scholars working in the Mathematical Sciences.

The Statistics 2011 Canada Conference was organized as part of decennial conferences held at Concordia University since 1971. Professor T. D. Dwivedi, who started this series of conferences with the support from local Statistics groups in Montreal and Ottawa, had been very much part of these conferences until he retired in 1997. He couldn't be with us at this conference as he passed away in December 2008. Holding a joint meeting of the Forum for Interdisciplinary Mathematics (FIM), with one of these decennial conferences had been the dream of Professor Dwivedi and Professor Satya Mishra, the latter having been President of FIM for two terms (1999-2004). Professor Mishra worked with me in the original planning of the joint conference while he was going under the treatment of his cancer. Alas, he was not to witness the success of the Conference; he died in October 2010.

The Conference attracted around 250 participants representing government, education, and industry from all over the globe. Current issues and topics in many fields dealing with applied statistics and mathematics were discussed. The presentations featured nine plenary speakers, 147 speakers in invited sessions, and 48 contributions in contributed paper and poster presentations.

It was gratifying to receive many compliments after the successful conclusion of the Conference. I would like to quote an e-mail from one of the plenary speakers to underscore the high quality of the talks and organization: "I would like to congratulate you and your team for organizing a very successful conference. The conference provided opportunities to meet and to hear many of the stalwarts in our profession, whom it may be otherwise difficult to meet and to talk to. I am thankful for the care and support you and your colleagues provided". The task of this magnitude required help from many people, and the organizers are thankful to all those who helped. I have a sense of pride organizing this successful conference on behalf of the Organizing Committee and the Scientific Program Committee along with the Student Volunteers.

The papers included in this volume have been handled by a team of Associate Editors and have been recommended for publication after a thorough review. I personally thank all the associate editors and referees for their comprehensive and timely reviews. Further, I am thankful to all the authors for submitting their papers and for their continued co-operation.

I would like to express my appreciation to Ms. Jane Venettacci, Administrator, Department of Mathematics and Statistics, Concordia University for coordinating the organization of the conference and helping at various stages of the production of this volume. I would also like to thank Mr. Yubing Zhai, Executive Editor, World Scientific Publishing Co. (USA) and Ms. Yue He, Desk Editor, World Scientific Publishing Co. (Singapore) for their flexible attitude and for coordinating this volume's timely production.

Finally, I would like to thank my wife, Prabha, who had not been able to fully share in my time for almost two years since the commencement of the planning of the Conference, and who provided unconditional support in freeing me for its organization.

I sincerely hope that this volume will be an important resource for the research community in mathematics and statistics.

Yogendra P. Chaubey   Concordia University
Mathematics and Statistics   Montreal, July 2012

CONTENTS

HYPOTHESIS ASSESSMENT USING THE BAYES FACTOR AND RELATIVE BELIEF RATIO

Z. BASKURT AND M. EVANS*

*Department of Statistics, University of Toronto,
Toronto, Ontario M5S 3G3, Canada*
* E-mail: mevans@utstat.utoronto.ca

The Bayes factor is commonly used for assessing the evidence for or against a given hypothesis $H_0 : \theta \in \Theta_0$, where Θ_0 is a subset of the parameter space. In this paper we discuss the Bayes factor and various issues associated with its use. A Bayes factor is seen to be intimately connected with a relative belief ratio which provides a somewhat simpler approach to assessing the evidence in favor of H_0. It is noted that, when there is a parameter of interest generating H_0, then a Bayes factor for H_0 can be defined as a limit and there is no need to introduce a discrete prior mass for Θ_0 or a prior within Θ_0. It is further noted that when a prior on Θ_0 does not correspond to a conditional prior induced by a parameter of interest generating H_0, then there is an inconsistency in prior assignments. This inconsistency can be avoided by choosing a parameter of interest that generates the hypothesis. A natural choice of a parameter of interest is given by a measure of distance of the model parameter from Θ_0. This leads to a Bayes factor for H_0 that is comparing the concentration of the posterior about Θ_0 with the concentration of the prior about Θ_0. The issue of calibrating a Bayes factor is also discussed and is seen to be equivalent to computing a posterior probability that measures the reliability of the evidence provided by the Bayes factor.

1. Introduction

Suppose we have the following ingredients available for a statistical problem: a statistical model $\{f_\theta : \theta \in \Theta\}$ given by a set of possible probability density functions with respect to a support measure μ on a sample space $\mathcal{X}$, a prior probability density π with respect to a support measure ν on the parameter space Θ, and the observed data $x \in \mathcal{X}$ which has been generated by one of the distributions in the model. Further suppose that our goal is to assess the hypothesis $H_0 : \theta \in \Theta_0$ where $\Theta_0 \subset \Theta$, namely, having observed x we want to assess the evidence that the true value of θ is in Θ_0.

A common way to approach this problem, based on the ingredients provided, is to compute the Bayes factor in favour of H_0. In fact we can only do this when $0 < \Pi(\Theta_0) < 1$ where Π is the prior probability measure. In this case the Bayes factor is given by

$$BF(H_0) = \frac{\Pi(\Theta_0 \mid x)}{1 - \Pi(\Theta_0 \mid x)} / \frac{\Pi(\Theta_0)}{1 - \Pi(\Theta_0)} \tag{1}$$

where $\Pi(\cdot \mid x)$ is the posterior probability measure. So $BF(H_0)$ measures the change from *a priori* to *a posteriori* in the odds in favour of H_0. If $BF(H_0) > 1$, then the data have lead to an increase in our beliefs that H_0 is true and so we have evidence in favour of H_0. If $BF(H_0) < 1$, then the data have lead to a decrease in our beliefs that H_0 is true and we have evidence against H_0.

Several questions and issues arise with the use of (1). First we ask why it is necessary to compare the odds in favour of H_0 rather than comparing the prior and posterior probabilities of H_0? Since this seems like a very natural way to make such a comparison we define the *relative belief ratio of H_0* as

$$RB(H_0) = \frac{\Pi(\Theta_0 \mid x)}{\Pi(\Theta_0)} \tag{2}$$

whenever $0 < \Pi(\Theta_0)$. Also $RB(H_0) > 1$ is evidence in favour of H_0 while $RB(H_0) < 1$ is evidence against H_0. Note that, as opposed to the Bayes factor, the relative belief ratio is defined when $\Pi(\Theta_0) = 1$ but we then have that $RB(H_0) = 1$ for every x and this is uninteresting. The relationship between (1) and (2) is given by $BF(H_0) = (1 - \Pi(\Theta_0))RB(H_0)/(1 - \Pi(\Theta_0)RB(H_0))$ so $BF(H_0)$ and $RB(H_0)$ are 1-1 increasing functions of each other for fixed $\Pi(\Theta_0)$ but otherwise are measuring change in belief on different scales.

This raises the second question associated with both (1) and (2). In particular, what do we do when $\Pi(\Theta_0) = 0$ simply because Θ_0 is a lower dimensional subset of Θ? Certainly we want to assess hypotheses that correspond to lower dimensional subsets. The most common solution to this problem is to follow Jeffreys[12,13] and modify Π to be the mixture $\Pi_\gamma = \gamma \Pi_0 + (1 - \gamma)\Pi$ where $\gamma \in (0, 1)$ and Π_0 is a probability measure concentrated on Θ_0. We then have that $\Pi_\gamma(\Theta_0) = \gamma$ and

$$BF(H_0) = m_0(x)/m(x) \tag{3}$$

where m_0 is the prior predictive density obtained from the model and Π_0, and m is the prior predictive density obtained from the model and Π. Also, we have that $RB(H_0) = m_0(x)/(\gamma m_0(x) + (1 - \gamma)m(x))$. While these calculations are formally correct, it is natural to ask if this approach is necessary as it does not seem reasonable that we should have to modify the prior Π simply because our hypothesis has $\Pi(\Theta_0) = 0$. In fact we will argue in Section 2 that this modification is often unnecessary as we can unambiguously define $BF(H_0)$ and $RB(H_0)$ by replacing Θ_0 on the right in (1) and (2) by a sequence of sets Θ_ϵ, where $\Pi(\Theta_\epsilon) > 0$ and $\Theta_\epsilon \downarrow \Theta_0$ as $\epsilon \downarrow 0$, and taking the limit. The limits obtained depend on how the sequence Θ_ϵ is chosen but this ambiguity disappears when H_0 is *generated* by a parameter of interest $\lambda = \Lambda(\theta)$ via $\Theta_0 = \Lambda^{-1}\{\lambda_0\}$ for some λ_0. In fact, when we have such a Λ then, using the definition via limits, $BF(H_0) = RB(H_0)$ and the two approaches to measuring change in belief are equivalent. Furthermore, if Π_0 is taken to be the conditional prior of θ given $\Lambda(\theta) = \lambda_0$, then $BF(H_0)$ defined as a limit equals (3) but this is not generally the case when Π_0 is chosen arbitrarily.

It is not always the case, however, that there is a parameter of interest generating H_0. In Section 3, we will argue that, in such a situation it is better that we choose such a Λ, as the introduction of γ and Π_0 can induce an inconsistency into the analysis. This inconsistency arises due to the fact that Π_0 does not necessarily arise from Π via conditioning. A natural choice is proposed in Section 3 where Λ is chosen so that $\Lambda(\theta)$ is a measure of the distance of θ from Θ_0. This is referred to as the *method of concentration* as the Bayes factor and relative belief ratio are now comparing the concentration of the posterior about Θ_0 with the concentration of the prior about Θ_0. If the data have lead to a greater concentration of the posterior distribution about Θ_0 than the prior, then this is naturally evidence in favour of H_0. This is dependent on the choice of the distance measure but now the conditional prior assignments on Θ_0 come from the prior Π.

A third issue is concerned with the calibration of $BF(H_0)$ or $RB(H_0)$, namely, when are these values large enough to provide convincing evidence in favour of H_0, or when are these values small enough to provide convincing evidence against H_0? In Section 4 we discuss some inequalities that hold for these quantities. For example, inequality (12) supports the interpretation that small values of $RB(H_0)$ provide evidence against H_0 while inequality (13) supports the interpretation that large values of $RB(H_0)$ provide evidence in favour of H_0. While these inequalities are *a priori*, the *a posteriori* probability (14) is a measure of the reliability of the evidence presented by $RB(H_0)$ given the specific data observed. In essence (14) is quantifying the uncertainty in the evidence presented by $RB(H_0)$. For if (14) is small when $RB(H_0)$ is large, then there is a large posterior probability that the true value of the parameter of interest has an even larger relative belief ratio than the hypothesized value and so $RB(H_0)$ cannot be taken to be reliable evidence in favour of H_0 being true. On the other hand if (14) is large when $RB(H_0)$ is large, then indeed we can take the value $RB(H_0)$ as reliable evidence in favour of H_0. When $RB(H_0)$ is small, then a small value of (14) indicates that this is reliable evidence against H_0 and conversely for a large value of (14), although inequality (15) shows that this latter case is not possible. We also address the issue of when evidence against H_0 corresponds to practically meaningful evidence in the sense of whether or not we have detected a meaningful deviation from H_0. Similar comments apply to $BF(H_0)$.

The Bayes factor has been extensively discussed in the statistical literature. For example, Kass and Raftery[14] and Robert, Chopin, and Rousseau[16] provide excellent surveys. Our attention here is restricted to the case where the prior Π is proper. O'Hagan[15] defines a fractional Bayes factor and Berger and Perrichi[2] define an intrinsic Bayes factor for the case of improper priors.

Overall our purpose in this paper is to survey some recent results on the Bayes factor, provide some new insights into the meaning and significance of these results and illustrate their application through some simple examples. References to much more involved applications to problems of practical interest are also provided. Much of the technical detail is suppressed in this paper and can be found in Baskurt and Evans[1].

2. General Definition

Suppose that Π is discrete and that there is a parameter of interest $\lambda = \Lambda(\theta)$ generating H_0 via $\Theta_0 = \Lambda^{-1}\{\lambda_0\}$ where $0 < \Pi(\Theta_0) < 1$. We consider a simple example to illustrate ideas.

Example 1. *Discrete uniform prior on two points.*

Suppose that $\{f_\theta : \theta \in \Theta\}$ is a family of probability functions where we have $\Theta = \{1/4, 1/2\}^2, \theta = (\theta_1, \theta_2), f_{(\theta_1,\theta_2)}(x,y)$ is the Binomial$(m, \theta_1) \times$ Binomial(n, θ_2) probability function, Π is uniform on $\Theta, \lambda = \Lambda(\theta) = \theta_1 - \theta_2$ and we wish to assess the hypothesis $\Lambda(\theta) = \lambda_0 = 0$. So $\Theta_0 = \Lambda^{-1}\{\lambda_0\} = \{(1/4, 1/4), (1/2, 1/2)\} = \{(\omega, \omega) : \omega \in \{1/4, 1/2\}\}$ and $\Pi(\Theta_0) = 1/2$.

Suppose we observe $x = 2, y = 2, m = 4$ and $n = 5$. Then $\Pi(\Theta_0 \,|\, x, y) = 0.512, BF(H_0) = (0.512)(0.5)/[(1 - 0.512)(0.5)] = 1.049$ and $RB(H_0) = 2(0.512) = 1.024$. So both the Bayes factor and the relative belief ratio provide marginal evidence in favour of H_0 as beliefs have only increased slightly after having seen the data.

When Π_Λ is discrete we can write $RB(H_0) = \pi_\Lambda(\lambda_0 \mid x)/\pi_\Lambda(\lambda_0)$ and $BF(H_0) = \pi_\Lambda(\lambda_0 \mid x)(1 - \pi_\Lambda(\lambda_0))/[\pi_\Lambda(\lambda_0)(1 - \pi_\Lambda(\lambda_0 \mid x))]$ where π_Λ and $\pi_\Lambda(\cdot \mid x)$ denote the prior and posterior probability functions of Λ. Also note that, in the discrete case, the Bayes factor and relative belief ratio are invariant to the choice of Λ generating H_0 via $\Theta_0 = \Lambda^{-1}\{\lambda_0\}$. So, for example, we could take $\Lambda(\theta) = I_{\Theta_0}(\theta)$ where I_{Θ_0} is the indicator function of Θ_0. When there is a particular Λ of interest generating H_0 via $\Theta_0 = \Lambda^{-1}\{\lambda_0\}$, we will write $BF(\lambda_0)$ and $RB(\lambda_0)$ for the Bayes factor and relative belief ratios for H_0, respectively.

When $\Pi(\Theta_0) = 0$ a problem arises as $BF(H_0)$ and $RB(H_0)$ are not defined. As discussed in Baskurt and Evans[1], however, when there is a parameter of interest generating H_0 via $\Theta_0 = \Lambda^{-1}\{\lambda_0\}$, then sensible definitions are obtained via limits of sets shrinking to λ_0. For this we need to assume a bit more mathematical structure for the problem as described in Baskurt and Evans[1]. With this, the marginal prior density of $\lambda = \Lambda(\theta)$ is given by

$$\pi_\Lambda(\lambda) = \int_{\Lambda^{-1}\{\lambda\}} \pi(\theta) J_\Lambda(\theta)\, \nu_{\Lambda^{-1}\{\lambda\}}(d\theta) \tag{4}$$

with respect to volume (Lebesgue) measure ν_Λ on the range of Λ, where

$$J_\Lambda(\theta) = \left(\det(d\Lambda(\theta))(d\Lambda(\theta))^t\right)^{-1/2}, \tag{5}$$

$d\Lambda$ is the differential of Λ and $\nu_{\Lambda^{-1}\{\lambda\}}$ is volume measure on $\Lambda^{-1}\{\lambda\}$. Furthermore, the conditional prior density of θ given $\Lambda(\theta) = \lambda$ is

$$\pi(\theta \mid \lambda) = \pi(\theta) J_\Lambda(\theta)/\pi_\Lambda(\lambda) \tag{6}$$

with respect to $\nu_{\Lambda^{-1}\{\lambda\}}$. The posterior density of $\lambda = \Lambda(\theta)$ is then given by

$$\pi_\Lambda(\lambda \mid x) = \int_{\Lambda^{-1}\{\lambda\}} \pi(\theta) f_\theta(x) J_\Lambda(\theta)\, \nu_{\Lambda^{-1}\{\lambda\}}(d\theta)/m(x). \tag{7}$$

Note that in the discrete case the support measures are counting measure and in the continuous case these are things like length, area, volume and higher dimensional analogs.

Now suppose that $C_\epsilon(\lambda_0)$ is a sequence of neighborhoods shrinking nicely to λ_0 as $\epsilon \to 0$ with $\Pi_\Lambda(C_\epsilon(\lambda_0)) > 0$ for each ϵ; see Rudin[18] for the technical definition of 'shrinking nicely'. Then

$$RB(C_\epsilon(\lambda_0)) \to RB(\lambda_0) = \pi_\Lambda(\lambda_0 \mid x)/\pi_\Lambda(\lambda_0)$$

and

$$BF(C_\epsilon(\lambda_0)) \to BF(\lambda_0) = \pi_\Lambda(\lambda_0 \mid x)/\pi_\Lambda(\lambda_0)$$

as $\epsilon \to 0$ where π_Λ and $\pi_\Lambda(\cdot \mid x)$ are now the prior and posterior densities of Λ with respect to ν_Λ. Note that $BF(\lambda_0) = RB(\lambda_0)$ when $\Pi(\Theta_0) = 0$. So the Bayes factor and relative belief ratio of H_0 are naturally defined as limits. Note that the limiting relative belief ratio takes the same form in the discrete and continuous case but this is not true for the Bayes factor.

An alternative expression for the limiting relative belief ratio is shown in Baskurt and Evans[1] to be

$$RB(\lambda_0) = m(x \mid \lambda_0)/m(x) \tag{8}$$

where $m(\cdot \mid \lambda_0)$ is the conditional prior predictive density of the data given $\Lambda(\theta) = \lambda_0$. The equality

$$m(x \mid \lambda_0)/m(x) = \pi_\Lambda(\lambda_0 \mid x)/\pi_\Lambda(\lambda_0) \tag{9}$$

is the Savage-Dickey ratio result and this holds for discrete and continuous models; see Dickey and Lientz[5] and Dickey[4]. Note that (8) is not a Bayes factor in the discrete case. We conclude that, when H_0 arises from a parameter of interest via $\Theta_0 = \Lambda^{-1}\{\lambda_0\}$, there is no need to introduce a discrete mass of prior probability on Θ_0 to obtain the Bayes factor in favour of H_0 and the conditional prior on Θ_0 is unambiguously given by $\pi(\cdot \mid \lambda_0)$.

We consider a simple application of this.

Example 2. *Continuous prior with Λ specified.*

Suppose $\{f_\theta : \theta \in \Theta\}$ is the family of distributions where $\Theta = [0,1]^2, \theta = (\theta_1, \theta_2), f_{(\theta_1, \theta_2)}(x, y)$ is the Binomial$(m, \theta_1) \times$ Binomial(n, θ_2) probability function, Π is uniform on $\Theta, \lambda = \Lambda(\theta) = \theta_1 - \theta_2$ and we wish to assess the hypothesis $\lambda = \lambda_0 = 0$. So $\Theta_0 = \Lambda^{-1}\{\lambda_0\} = \{(\theta, \theta) : \theta \in [0,1]\}$ and $\Pi(\Theta_0) = 0$. Using (5) we have $J_\Lambda(\theta) = 1/\sqrt{2}$ and, since $\nu_{\Lambda^{-1}\{\lambda\}}$ is length measure on $\Lambda^{-1}\{\lambda\}$, applying this to (4) gives $\pi_\Lambda(\lambda) = 1 - \lambda$ when $\lambda \geq 0, \pi_\Lambda(\lambda) = 1 + \lambda$ when $\lambda \leq 0$ and $\pi_\Lambda(\lambda) = 0$ otherwise. To compute $RB(\lambda) = \pi_\Lambda(\lambda \mid x, y)/\pi_\Lambda(\lambda)$ for a general Λ, we will typically have to sample from the prior to obtain π_Λ but here we have an exact expression for the prior. From (6) we see that the conditional prior density of θ given $\lambda = \Lambda(\theta)$, with respect to length measure on $\{\theta : \lambda = \Lambda(\theta)\}$, is given by $\pi(\theta \mid \lambda) = 1/\sqrt{2}(1 - \lambda)$ when $\lambda \geq 0$ and by $\pi(\theta \mid \lambda) = 1/\sqrt{2}(1 + \lambda)$ when $\lambda \leq 0$. Therefore, the conditional prior is uniform.

The posterior distribution of θ is given by

$$(\theta_1, \theta_2) \mid (x, y) \sim \text{Beta}(x + 1, m - x + 1) \times \text{Beta}(y + 1, n - y + 1).$$

Suppose we observe $x = 2, y = 2, m = 4$ and $n = 5$. In Figure 1 we present a plot of the posterior density of $\lambda = \Lambda(\theta)$ and in Figure 2 a plot of the

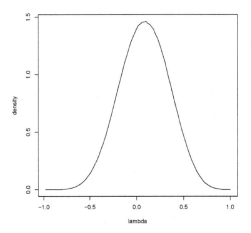

Fig. 1. Plot of estimate of the posterior density of $\lambda = \Lambda(\theta)$ in Example 2.

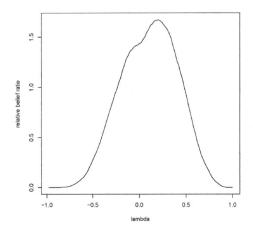

Fig. 2. Plot of estimate of the relative belief ratios of $\lambda = \Lambda(\theta)$ in Example 2.

relative belief ratio $RB(\lambda)$ as a function of λ based on samples of 10^5 from the prior and posterior. The maximum value of $RB(\lambda)$ is 1.68 and this occurs at $\hat{\lambda} = 0.22$, so $RB(\hat{\lambda}) = 1.68$. The relative belief ratio in favour of H_0 is given by $RB(0) = 1.47$ and so we have evidence in favour of H_0. This evidence doesn't seem overwhelming, although it does say that the data has lead to an approximate 50% increase in the "probability" of Θ_0.

3. The Method of Concentration

It is a feature of many problems, however, that H_0 does not arise from an obvious parameter of interest $\lambda = \Lambda(\theta)$ and so appropriate definitions of $BF(H_0)$ and $RB(H_0)$ are ambiguous when $\Pi(\Theta_0) = 0$.

Example 3. *Continuous prior with Λ unspecified.*

Suppose that the situation is identical to that described in Example 2 but now we do not declare that we want to make inference about $\lambda = \Lambda(\theta) = \theta_1 - \theta_2$. Rather it is simply specified that we wish to assess the hypothesis $\Theta_0 = \{(\omega, \omega) : \omega \in [0, 1]\}$, namely, we only specify that we are interested in determining whether or not $\theta_1 = \theta_2$. Of course, $\Theta_0 = \Lambda^{-1}\{0\}$ but there are many such Λ generating H_0 and it is not clear, given the statement of the problem, which we should use. It seems clear from (4) and (5), however, that $RB(H_0) = RB(\lambda_0)$ will depend on the Λ we use to obtain the relative belief ratio.

It is common practice in such situations to follow Jeffreys and replace Π by the prior $\Pi_\gamma = \gamma\Pi_0 + (1 - \gamma)\Pi$ to compute $BF(H_0)$ and $RB(H_0)$. From (3) we see that, when Π_0 corresponds to the conditional prior of $\lambda = \Lambda(\theta)$ given that $\lambda = \lambda_0$, then $m_0(x) = m(x \mid \lambda_0)$ and (3) equals (8). Again there is no need to introduce γ.

On the other hand, if Π_0 is not equal to the conditional prior based on Π for *some* Λ, then a fundamental inconsistency arises in the Bayesian analysis, as the conditional beliefs on Θ_0 do not arise from the prior Π. The existence of such an inconsistency when Π is discrete is unacceptable and there is no reason to allow this in the continuous case either.

Example 4. *Discrete uniform prior on many points.*

Consider the context of Example 1 where now $\Theta = \{1/k, 2/k, \ldots, (k - 2)/(k - 1)\}^2$ for some positive integer k and Π is the uniform prior on this set. Suppose again we want to assess the hypothesis $\Theta_0 = \{(\omega, \omega) : \omega \in 1/k, 2/k, \ldots, (k - 2)/(k - 1)\}\}$. The hypothesis being assessed is clearly the assertion that $\theta_1 = \theta_2$. Note that in this case, however we choose Λ generating H_0, the relative belief ratio in favour of H_0 is the same and the conditional prior given Θ_0 is unambiguously the uniform prior on Θ_0. Furthermore, when k is very large we can think of the continuous problem in Example 3 as an approximation to this problem. Without specifying a Λ in Example 3, however, we are not specifying how this approximation is taking place.

Avoiding the inconsistency requires the choice of a suitable Λ that generates H_0 via $\Theta_0 = \Lambda^{-1}\{\lambda_0\}$ and then using the relative belief ratio $RB(\lambda_0)$. Faced with the option of either choosing γ and Π_0 or choosing a Λ that generates H_0, the latter seems preferable as it ensures that beliefs are being assigned consistently in the problem.

The effect of the inconsistency can be seen directly via a generalization of the Savage-Dickey ratio result (9) due to Verdinelli and Wasserman[19] and that is also discussed in Baskurt and Evans[1]. This result was derived to aid in the computation of (3) but in fact it relates the Bayes factor obtained using the Jeffreys approach via γ and Π_0 and that obtained via the definition in Section 2 as a limit. This can be written in two ways as

$$m_0(x)/m(x) = RB(\lambda_0)E_{\Pi_0}\left(\pi(\theta \,|\, \lambda_0, x)/\pi(\theta \,|\, \lambda_0)\right)$$
$$= RB(\lambda_0)E_{\Pi(\cdot \,|\, \lambda_0, x)}\left(\pi_0(\theta)/\pi(\theta \,|\, \lambda_0)\right). \tag{10}$$

The first equality in (10) says that (3) equals the Bayes factor obtained as a limit in Section 2, times the expected conditional relative belief ratio of θ computed as a limit via $\Pi(\cdot \,|\, \lambda_0)$, where the expectation is taken with respect to the prior Π_0 placed on Θ_0. It is interesting to note that the adjustment factor involves relative belief ratios. The second equality in (10) says that (3) equals the Bayes factor obtained as a limit in Section 2, times the expected ratio of the prior π_0 evaluated at θ to the conditional prior induced by Π evaluated at θ, given that $\lambda_0 = \Lambda(\theta)$, where the expectation is taken with respect to the conditional posterior induced by the prior Π, given that $\lambda_0 = \Lambda(\theta)$. So if π_0 is very different from any conditional prior $\pi(\cdot \,|\, \lambda_0)$ induced by Π, then we can expect a big differences in the Bayes factors obtained by the two approaches.

To avoid this inconsistency a natural choice of Λ in such a problem is to take $\Lambda = d_{\Theta_0}$, where $d_{\Theta_0}(\theta)$ is a measure of the distance θ is from Θ_0. Therefore, $d_{\Theta_0}(\theta) = 0$ if and only if $\theta \in \Theta_0$ and $\Theta_0 = \Lambda^{-1}\{0\}$. With this choice the Bayes factor (and relative belief ratio) $RB(0)$ is a comparison of the concentration of the posterior distribution about Θ_0 with the concentration of the prior distribution about Θ_0. If $RB(0) > 1$, then the posterior distribution is concentrating more about Θ_0 than the prior and we have evidence in favour of H_0. This seems quite natural as when H_0 is true we expect the posterior distribution to assign more of its mass near Θ_0 than the prior, otherwise the data is providing evidence against H_0. Under weak conditions $RB(0)$ will converge to 0 when H_0 is false and to ∞ when H_0 is true, as the amount of data increases.

Of course, there is still an arbitrariness as there are many possible choices of distance measure. But this arbitrariness is in essence an unavoidable consequence of a problem that is not fully specified. The problem is similar to the Borel paradox in probability theory where, in general, there is no unique conditional probability measure associated with conditioning on a set of measure 0 even though this may make perfect sense as in Example 2. The way out of this is to specify a function that generates the set but the conditional distribution depends on the function chosen. Our recommendation is that we should address the problem in a way that guarantees beliefs assigned in a consistent way and this effectively entails choosing a Λ that generates H_0. If we choose Λ in an intuitively satisfying way then this adds greater support for this approach. Setting $\Lambda = d_{\Theta_0}$ for some distance measure d_{Θ_0} satisfies these criteria.

In a number of problems, see the discussion in Section 5, we have chosen d_{Θ_0} to be squared Euclidean distance so we are effectively using least squares. This choice often exhibits a very nice property as expressed in the following result.

Proposition 1. $RB(\lambda_0)$ is the same for all Λ in the set $\{\Lambda : J_\Lambda(\theta)$ is constant and nonzero for all $\theta \in \Lambda^{-1}\{\lambda_0\}\}$.

Proof: From (4) and (7) we have that

$$RB(\lambda_0) = \frac{\pi_\Lambda(\lambda_0 \mid x)}{\pi_\Lambda(\lambda_0)} = \frac{\int_{\Lambda^{-1}\{\lambda_0\}} \pi(\theta) f_\theta(x) J_\Lambda(\theta) \, \nu_{\Lambda^{-1}\{\lambda_0\}}(d\theta)/m(x)}{\int_{\Lambda^{-1}\{\lambda_0\}} \pi(\theta) J_\Lambda(\theta) \, \nu_{\Lambda^{-1}\{\lambda_0\}}(d\theta)}$$

$$= \frac{\int_{\Theta_0} \pi(\theta) f_\theta(x) \, \nu_{\Theta_0}(d\theta)/m(x)}{\int_{\Theta_0} \pi(\theta) \, \nu_{\Theta_0}(d\theta)}$$

where the last equality follows from $\Theta_0 = \Lambda^{-1}\{\lambda_0\}$ and $\nu_{\Lambda^{-1}\{\lambda_0\}} = \nu_{\Theta_0}$ since this measure is determined by the geometry of Θ_0 alone. $\square$

As already noted, Proposition 1 always holds when Π is discrete because the parameter space is countable and with the discrete topology on Θ, all functions are continuously differentiable with $J_\Lambda(\theta) \equiv 1$. Also, whenever $\Theta_0 = \{\theta_0\}$ for some $\theta_0 \in \Theta$, then any Λ continuously differentiable at θ_0 that generates $\Theta_0 = \{\theta_0\}$ clearly satisfies the condition of Proposition 1.

When $\Theta_0 = \mathcal{L}(A)$ for some $A \in R^{k \times m}$ of rank m and $\Lambda(\theta) = d_{\Theta_0}(\theta) = ||\theta - (A'A)^{-1}A'\theta||^2$ is the squared Euclidean distance of θ from Θ_0, then $J_\Lambda(\theta) = ||\theta - (A'A)^{-1}A'\theta||^{-1}/2 = \Lambda^{-1/2}(\theta)/2$ and so if $\lambda_0 > 0$ the conditions of Proposition 1 are satisfied. When $\lambda_0 \to 0$ we have that

$$RB(\lambda_0) \to RB(0) = \frac{\int_{\Theta_0} \pi(\theta) f_\theta(x) \, \nu_{\Theta_0}(d\theta)/m(x)}{\int_{\Theta_0} \pi(\theta) \, \nu_{\Theta_0}(d\theta)} \tag{11}$$

which is independent of Λ. So when Λ generating H_0 satisfies the requirement of Proposition 1, we see that volume distortions induced by Λ, as measured by $J_\Lambda(\theta)$, do not affect the value of the relative belief ratio.

We consider an example using squared Euclidean distance.

Example 5. *Continuous prior using concentration.*

Suppose the situation is as in Example 3. We take $\Lambda_*(\theta) = d_{\Theta_0}(\theta)$ to be the squared Euclidean distance of θ from $\Theta_0 = \{(\omega, \omega) : \omega \in [0, 1]\}$. It is clear that $\Lambda_*(\theta) = (\theta_1 - \theta_2)^2/2$, $J_{\Lambda_*}(\theta) = \Lambda_*^{-1/2}(\theta)/2$ and (11) applies. Now notice that the Λ used in Example 2 also satisfies the conditions of Proposition 1 and so $\lim_{\lambda_0 \to 0} RB(\lambda_0) = RB(0)$ must be the same as what we get when using Λ_* and (11). Therefore, for the data as recorded in Example 2 we also have $RB(0) = 1.47$ when using the method of concentration.

The outcome in Example 5 is characteristic of many situations when taking $d_{\Theta_0}(\theta)$ to be the squared Euclidean distance of θ from Θ_0.

4. Calibration

In Baskurt and Evans[1] several *a priori* inequalities were derived that support the interpretations of $BF(\lambda_0)$ or $RB(\lambda_0)$ as evidence for or against H_0. These are generalizations to the Bayesian context of inequalities derived in Royall[17] for likelihood inferences. For example, it can be proved that

$$M\left(m(X \mid \lambda_0)/m(X) \leq RB(\lambda_0) \mid \lambda_0\right) \leq RB(\lambda_0) \tag{12}$$

and

$$M(\cdot \mid \lambda_0) \times \Pi_\Lambda\left(m(X \mid \lambda)/m(X) \geq RB(\lambda_0)\right) \leq (RB(\lambda_0))^{-1}. \tag{13}$$

In both inequalities we consider $RB(\lambda_0)$ as a fixed observed value of the relative belief ratio and $X \sim M(\cdot \mid \lambda_0)$ where $M(\cdot \mid \lambda_0)$ is the conditional prior predictive measure given that $\Lambda(\theta) = \lambda_0$. So inequality (12) says that the conditional prior probability, given that H_0 is true, of obtaining a value of the relative belief ratio (recall (9)) of H_0 smaller than the observed value is bounded above by $RB(\lambda_0)$. So if $RB(\lambda_0)$ is very small this probability is also very small and we can consider a small value of $RB(\lambda_0)$ as evidence against H_0. In (13) $\lambda \sim \Pi_\Lambda$ independent of X and the inequality says that the conditional prior probability, given that H_0 is true, of obtaining a larger value of the relative belief ratio at a value λ generated independently from Π_Λ, is bounded above by $(RB(\lambda_0))^{-1}$. So if $RB(\lambda_0)$ is very large then, when H_0 is true, it is extremely unlikely that we would obtain a larger value of $RB(\lambda)$ at a value λ that is *a priori* reasonable. So we can consider large values of $RB(\lambda_0)$ as evidence in favour of H_0. Note that these inequalities also apply to the Bayes factor in the continuous case and similar inequalities can be derived for the Bayes factor in the discrete case.

We now consider the reliability or the uncertainty in the evidence given by $RB(\lambda_0)$. We will measure the reliability of this evidence by comparing it to the evidence in favour of alternative values of λ. For, if $RB(\lambda_0)$ is very large, so we have strong evidence in favour of H_0, but $RB(\lambda)$ is even larger for values of $\lambda \neq \lambda_0$, then this casts doubt on the reliability of the evidence in favour of H_0. The probabilities in (12) and (13) are *a priori* measures of the reliability of the evidence given by $RB(\lambda_0)$. For inequality (12) tells us that, when $RB(\lambda_0)$ is very small, there is little prior probability of getting an even smaller value of this quantity when H_0 is true. Similarly, (13) tells us that, when $RB(\lambda_0)$ is very large, there is little prior probability of getting an even larger value for $RB(\lambda)$ for some λ when H_0 is true. Based upon fundamental considerations, however, we know that we need to measure the reliability using posterior probabilities. Accordingly, we propose to use the posterior tail probability

$$\Pi\left(RB(\Lambda(\theta)) \leq RB(\lambda_0) \,|\, x\right) \tag{14}$$

to measure the reliability, or equivalently quantify the uncertainty, in the evidence given by $RB(\lambda_0)$. We see that (14) is the posterior probability of a relative belief ratio (and Bayes factor in the continuous case) $RB(\lambda)$ being no larger than $RB(\lambda_0)$. If $RB(\lambda_0)$ is large and (14) is small (see Baskurt and Evans[1] for examples where this occurs), then evidence in favour of H_0, as expressed via $RB(\lambda_0)$, needs to be qualified by the fact that our posterior beliefs are pointing to values of λ where the data have lead to an even bigger increase in belief. In such a situation the evidence in favour of H_0 does not seem very reliable. If $RB(\lambda_0)$ is large and (14) is also large, then we have reliable evidence in favour of H_0. Similarly, if $RB(\lambda_0)$ is small and (14) is small, then we have reliable evidence against H_0, while a large value of (14) suggests the evidence against H_0, as expressed through $RB(\lambda_0)$, is not very reliable. In fact, the *a posteriori* inequality

$$\Pi\left(RB(\Lambda(\theta)) \leq RB(\lambda_0) \,|\, x\right) \leq RB(\lambda_0) \tag{15}$$

is established in Baskurt and Evans[1]. Inequality (15) shows that a very small value of $RB(\lambda_0)$ is always reliable evidence against H_0.

We note that (14) is not interpreted like a P-value. The evidence for or against H_0 is expressed via $RB(\lambda_0)$ and (14) is a measure of the reliability of this evidence. A similar inequality exists for the Bayes factor in the discrete case.

It is a substantial advantage of the use of the Bayes factor and the relative belief ratio that they can express evidence both for and against hypotheses. A weakness of this approach is that somewhat arbitrary scales have been created for comparison purposes for the Bayes factor. For example, according to such a scale a Bayes factor of 20 is considerable evidence in favour of λ_0. But, as shown in Baskurt and Evans[1], it is possible that there are other values of λ for which the Bayes factor is even larger and this leads one to doubt the evidence in favour of λ_0. If, however, such values have only a small amount of posterior weight, then the evidence in favour of λ_0 seems more compelling. Assessing this is the role of (14).

As with the use of P-values, however, we need to also take into account the concept of practical significance when assessing hypotheses. It is clear that $RB(\lambda_0)$, even together with (14), doesn't do this. For example, suppose $RB(\lambda_0)$ is small and (14) is small so we have evidence against H_0, or suppose $RB(\lambda_0)$ is large while (14) is small or at least not large. In such a situation it is natural to compute $(\hat{\lambda}, RB(\hat{\lambda}))$ where $\hat{\lambda} = \arg\sup RB(\lambda)$ as $\hat{\lambda}$ is in a sense the value of λ best supported by the data. If the difference between $\hat{\lambda}$ and λ_0 is not of practical significance, then it seems reasonable to proceed as if H_0 is true. So the approach discussed here can also take into account the notion of practical significance.

We consider an example involving computing (14).

Example 6. *Measuring the reliability of a Bayes factor.*

Consider the context discussed in Example 2. We had $\lambda = \Lambda(\theta) = \theta_1 - \theta_2$ and we wished to assess the hypothesis $\Lambda(\theta) = \lambda_0 = 0$, so $\Theta_0 = \Lambda^{-1}\{\lambda_0\} = \{(\theta, \theta) : \theta \in [0, 1]\}$. For the data specified there, we obtained $RB(0) = 1.47$ and we have evidence in favour of H_0 although not overwhelmingly strong. Using simulation we computed (14) as equal to 0.42. This indicates that the posterior probability of obtaining a larger value of the Bayes factor is 0.58 so the evidence in favour of H_0 is not particularly reliable. Of course we have very small samples here so this is not surprising.

The maximum value of $RB(\lambda)$ is 1.68 and this occurs at $\hat{\lambda} = 0.22$, so $RB(\hat{\lambda}) = 1.68$. If in the particular application we also thought that $\hat{\lambda} = 0.22$ does not differ meaningfully from 0, then it seems reasonable to proceed as if H_0 is true.

5. Conclusions

This paper has shown that is possible to provide a sensible definition of a Bayes factor in favour of a hypothesis $H_0 : \theta \in \Theta_0$, in contexts where Θ_0 has prior probability 0 simply because it is a lower dimensional subset of Θ and the prior Π is continuous on Θ. This is accomplished without the need to introduce the mixture prior $\Pi_\gamma = \gamma\Pi_0 + (1 - \gamma)\Pi$ which requires the specification of γ and Π_0, in addition to Π. Our approach avoids the need for the discrete mass γ but more importantly, when Π_0 is not given by the conditional prior of θ given that $\Lambda(\theta) = \lambda_0$ for some Λ generating H_0, it avoids an inconsistency in the assignment of prior beliefs. This provides a unified approach to Bayesian inference as we no longer have to treat estimation and hypothesis assessment problems separately, namely, we do not have to modify an elicited prior Π just to deal with a hypothesis assessment. In situations where H_0 is not generated by a parameter of interest, then the method of concentration is seen to be a natural approach to choosing such a Λ. We have also discussed the calibration of a Bayes factor or relative belief ratio and have shown that this is intimately connected with measuring the reliability of the evidence presented by a Bayes factor or relative belief ratio. Finally we have shown how a maximized Bayes factor or relative belief ratio can be used to assess the practical significance of the evidence presented by these quantities. While we have illustrated these ideas via simple examples in this paper much more substantive and practical applications can be found in Evans, Gilula and Guttman[6] and Evans, Gilula, Guttman and Swartz,[8] Cao, Evans and Guttman,[3] Evans, Gilula and Guttman[7] and Baskurt and Evans.[1]

Other inferences are closely related to those discussed in this paper. For example when we have a parameter of interest Λ, then the value $\hat{\lambda}$ maximizing $RB(\lambda)$, referred to as the *least relative surprise estimator (LRSE)*, can be used to estimate λ. It is shown in Evans and Jang[10] that $\hat{\lambda}$ is either a Bayes rule or a limit of Bayes rules where the losses are derived from the prior. A γ-credible region for λ is given by $C_\gamma(x) = \{\lambda_0 : \Pi\left(RB(\Lambda(\theta)) \leq RB(\lambda_0)\,|\,x\right) > 1 - \gamma\}$ and is referred to as a γ-*relative surprise region*. A variety of optimality properties have been established for these regions, when compared to other rules for the construction of Bayesian credible regions, in Evans, Guttman and Swartz[9], and Evans and Shakhatreh.[11]

References

1. Z. Baskurt and M. Evans, *Technical Report, 1105*, Department of Statistics, University of Toronto (2011).
2. J. O. Berger and R. L. Perrichi, *J. Amer. Stat. Assoc.* **91**, 10 (1996).
3. Y. Cao, M. Evans and I. Guttman, *Technical Report, 1003*, Department of Statistics, University of Toronto (2010).
4. J. M. Dickey, (1971). *Ann. Stat.* **42**, 204 (1971).
5. J. M. Dickey and B. P. Lientz, *Ann. Math. Stat.* **41**, 214 (1970).
6. M. Evans, Z. Gilula and I. Guttman, *Statistica Sinica* **3** 391, (1993).

7. M. Evans, Z. Gilula and I. Guttman, *Quant. Mark. and Econ.* (2012), to appear. DOI: 10.1007/s11129-011-9116-1

8. M. Evans, Z. Gilula, I. Guttman and T. Swartz, *J. Amer. Stat. Assoc.* **92**, 208 (1997).

9. M. Evans, I. Guttman and T. Swartz, *Can. J. of Stat.* **34**, 113 (2006).

10. M. Evans and G.-H. Jang, *Technical Report, 1104*, Department of Statistics, University of Toronto (2011).

11. M. Evans and M. Shakhatreh, *Elec. J. of Stat.* **2**, 1268 (2008).

12. H. Jeffreys, *Proc. Camb. Phil. Soc.* **31**, 203 (1935).

13. H. Jeffreys, *Theory of Probability,* 3rd Edition (Oxford University Press, Oxford, U.K., 1961).

14. Kass, R.E. and Raftery, A.E. (1995) *J. Amer. Statist. Assoc.* **90**, 773 (1995).

15. O'Hagan, A. (1995) *J. Royal Statist. Soc.* **B56**, 3 (1995).

16. C. P. Robert N. Chopin and J. Rousseau, *Statist. Sci.* **24** 141 (2009).

17. R. Royall, *J. Amer. Statist. Assoc.* **95**, 760 (2000).

18. W. Rudin, *Real and Complex Analysis*, 2nd Edition (McGraw-Hill, New York, 1974).

19. I. Verdinelli and L. Wasserman, *J. Amer. Stat. Assoc.* **90**, 614 (1995).

AN ANALYSIS OF 1990-2011 ONTARIO SURFACE AIR TEMPERATURES

DAVID R. BRILLINGER[a]

*Statistics Department, University of California, 367 Evans,
Berkeley, CA 94720-3860, USA*

E-mail: brill@stat.berkeley.edu

The paper's concern is the estimation of the average monthly temperature across a given region as a function of time. The region studied here is the Canadian province of Ontario and the time unit is month. Data for various stations and times were obtained from the Berkeley Earth website, (http://www.berkeleyearth.org). In this paper a generalized additive model with random effects is employed that allows both spatial and temporal dependence. Handling variability in both space and time is basic.

1. Introduction

This study uses historical data to analyse Earth's surface temperatures as functions of time, t, and space, (x,y). The particular focus is the monthly averages for the Canadian province of Ontario. These particular data evidence a wide variety of the complications common to surface temperature studies. The data employed were obtained from the Berkeley Earth project[2,12,13].

The temperatures are measured at stations in a regular temporal fashion and typically there are many temperature values at the same (x,y). However these locations are distributed irregularly. Further, stations enter and leave as time passes so there are starts and stops and gaps in the temporal sequences. This may be seen in the right hand panel of Fig. 1 below where the measurement times are shown as related to their latitude. The left hand pane provides the geographic locations of the stations. One sees that the measurements get scarcer as one moves north in the province.

Biases may be anticipated to be present in the results of equally weighted analyses if the spatial and temporal irregularities are not dealt with. The temporal

---

[a] Work partially supported by NSF grant DMS-100707157.

case will be handled here, for now, by restricting consideration to stations not missing too many values. The spatial aspect will be dealt with by introducing weights, a continuous form of post-stratification. Generalized additive models with random effects are employed.

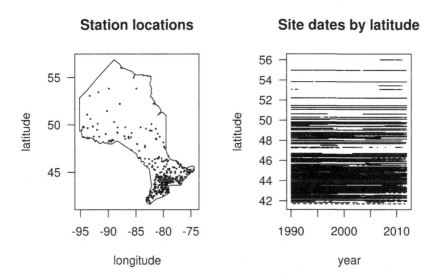

Station locations   **Site dates by latitude**

Fig. 1. The left panel show the locations of 1990-2011 stations. The right panel highlights the stations' dates and related latitude.

The statistical package R,[11] is employed in the analyses throughout. The R-libraries employed included mgcv, spatialkernel, RandomFields and splancs. The library mgcv works with regression splines. It includes smoothing parameter estimation by cross-validation.

2. The Data

Fig. 1 may be interpreted in general terms. Consider the left hand panel. The bottom and left hand side border on water, the Lakes Huron, Erie and Ontario. As one moves north in the province the temperature gets cooler. The winters and summers can both be extremely hot. The circles in the plot are the locations of the stations that existed at some time in the period 1990-2011. There are 440 of these. The panel shows a much greater density of stations in the south. The right panel is consistent with this. In addition it shows the temporal gaps in coverage of the province generally and the station monthly values specifically.

For this time period there are 264 data months, however December 2011 was missing throughout. A decision was taken to exclude any station missing more than 20 months at this stage.

To deal with the remaining missing values an elementary analysis of variance model was employed, specifically

$$Y_{jk} = \alpha_j + \beta_k + \varepsilon_{jk} \tag{1}$$

with j indexing years and k months that was fitted by least squares. This provided estimates of the December 2011 values in particular. Here Y is temperature, α a year effect, β a month effect and ε is the error term. The locations remained irregularly spread, as may be seen in Fig. 4 to come. The effects of this will be ameliorated by employing weights, a form of continuous post-stratification. The fit was by simple least squares and the estimates $a_{j\_} + b_k$ then employed as needed. This led to a complete array of 264 by 81 "temperatures". The 81 cases will be referred to as the basic subset of stations. The 264 time points were all spaced 1 month apart.

The data for one of the stations, Toronto Centre, is provided in the top panel of Fig. 2. The dominant feature in the graph is the annual 12 month seasonal. It provides the vast majority of the variability in the series. The monthly average temperature may be seen to range from about -25 degrees centigrade to about +20 during the period 1990 to November 2011.

3. Models and Preliminary Analyses

The model (1) provided an elementary method to infer missing values. Often one seeks regular arrays of data as various statistical methods have such in mind, however more complex methods could be employed. The missing time observations were dealt with from the start. Fig. 1 right hand panel shows the importance of this. At the outset stations with more than 20 months missing were excluded

As an illustrative analysis we consider the 1990 to 2011 data for Toronto Centre with the December 2011 value filled in. Next consider the following stochastic model

$$Y(t) = \alpha(\text{year}) + \beta(\text{mon}) + \varepsilon(t), \tag{2}$$

where year and mon (month) are factors and $\varepsilon(.)$ is a stationary autoregressive series of order 1. This is the model Wood[14] considers in Section 6.7.2 of his book. He provides R-commands and results for an analysis of daily Cairo, Egypt temperature for a stretch of 3780 days of data starting 1 January 1995. He employs the R-library mgcv and presents a variety of results. The fitting procedures employed are based on an assumption of Gaussian ε. The series $\{\varepsilon(t)\}$ is assumed to be a stationary AR(1) and is treated as an additive random effect in the model. The functions α and β are assumed to be smooth. The actual fitting is via the R-function gamm(.) employing maximum likelihood. The functions α and β are represented as splines. Estimation of numbers and locations of knots is included in the function gamm(.).

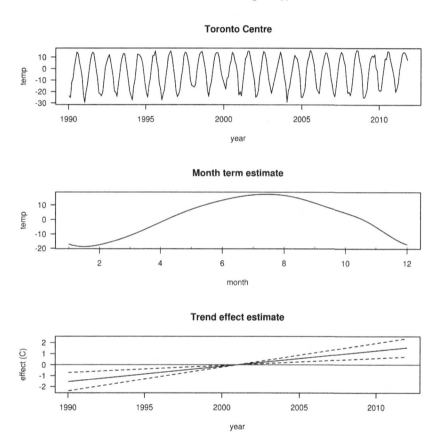

Fig. 2. Top panel: Toronto Centre monthly temperatures. Middle panel: Estimated seasonal effect Bottom panel: Toronto trend effect estimate and associated uncertainties.

Some results of carrying out the fitting of model (2) are shown in Fig. 2. The top panel, already discussed, provides the data themselves. The seasonal effect is clear and meant to be handled by β. The middle panel provides its estimate. That effect is well known and understood to a degree. The quantity of topical interest is α where climate change might lurk. It is interesting that the estimate of α here is approximately linear in time. The bowtie lines in the figure are one way to display 2 standard error variations. It has in mind prediction purposes. One of the outer limits of the bounds is almost on top of the zero line. There remain the data from the other stations to possibly clarify the situation.

Station trend effects

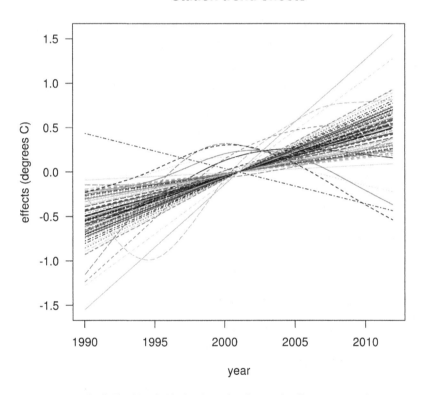

Fig. 3. The 81 individual estimated station trends effects superposed.

4. Results

In the next analyses trend effect terms are estimated for each of the 81 stations and then the trends merged. These trend curves are plotted in Fig. 3. The curve for Toronto Centre appearing in Fig. 2 is one of them. There is considerable scatter but a central core of curves heading from the lower left to the upper right.

To make inferences one needs measures of uncertainty and these need to take note of temporal and spatial variability. In the figure, one sees outliers highlighting the need for a robust/resistant method of merging the 81 curves.

Weights - subset of 81 stations

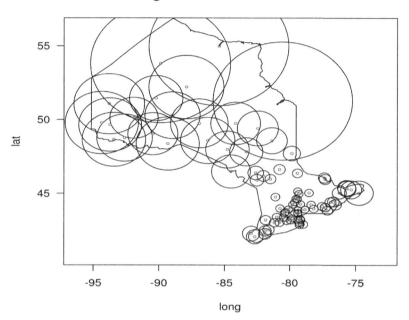

Fig. 4. The weights employed in the merging work. There is one for each of the 81 stations station. The diameter of the circle is proportional to the density at the station. Also shown are the locations of the observation stations in Ontario.

The next step is to average the values at each time point, however spatial bias is a serious concern. One debiasing approach is to weight the stations in a region inversely to the density of stations there, Pertinent weights were determined as follows: the locations of the stations are viewed as a planar point process and ts intensity function estimated by kernel smoothing. This was realized by the function lambdahat of the R-library spatialkernel. If the estimated density at the location (x,y) is $d(x,y)$ the weight at (x,y) is $w(x,y) = 1/ d(x,y)$. For the data concerned the array is not changing for the time period of interest. The rates are displayed in Fig. 4 via the diameter of circles which are centered at the various array locations. As anticipated the weights get larger as one moves to the north of the province where stations are scarce.

For the moment a simple weighted average of the 81 estimated trend effects is employed. The weights are those of Fig. 4 and the resulting curve is the central one in Fig. 5. Assuming the 81 values at a given time point are statistically independent there is the classic formula for the variance of a weighted mean and an estimate for it. The bounding curves in the figure are the approximate ± 2 *s.e.* limits.

Merged trend effect

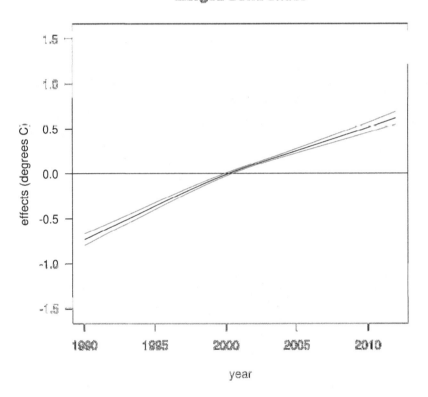

Fig. 5. Individual station trend effects with weights inversely proportional to station density. There are 2 *s.e.* bounds.

Looking at Fig. 5 it is hard not to infer that there is an increasing trend for this province and time period. However serial dependence has to be a concern as it may be affecting the estimated standard error. Perhaps the AR(1) didn't go far enough. However there is a multiplier that may be motivated and applied to standard errors after a least squares fit to take some note of remaining temporal dependence. In the present case it is estimated as 1.4015. The source of this number is the average of the first 10 periodogram values divided by the average of all; see for Example p. 127 in Hannan[8]. Including it will not change Fig. 5 a great deal.

5. Model Assessment

The R-library mgcv has a function that provides results for some common methods of assessing fit, in the case of independent observations. It leads to Fig. 6 here.

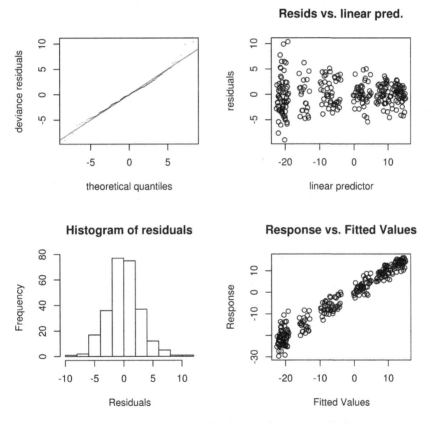

Fig. 6. The output produced by the mgcv function gam.check.

On examining these statistics, the marginal Gaussianity is suggested in the two left hand panels. This is important because the fitting and inference techniques employed are motivated by that assumption. The bottom right hand panel suggested that the responses are approximately linearly related to the fitted values. That would relate to how much of the variability of the temperatures was due to seasonal variability. The top right panel gives reason for further research. It reflects that the variability around the fit is greater at the low temperatures. One could deal with this by introducing further weights, but that is for future research.

Autocorrelation in the series ε of model (2) needs to be studied. The assumption of the independence of the error term, ε for the curves in Fig. 3 entered into the computation of the standard error bounds. Even though these bounds are marginal temporal dependence is a concern in their derivation.

log periodograms

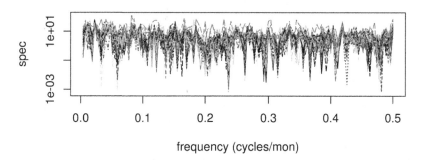

log mean periodogram

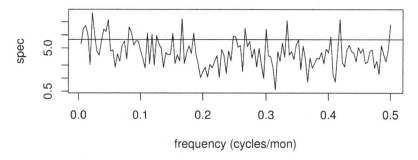

Fig. 7. The top panel shows 81 periodogram replicates. The lower provides the average of these. Both figures are graphed on "log paper".

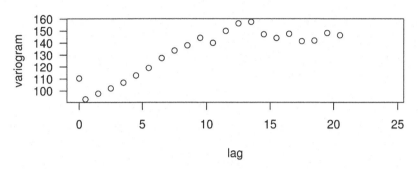

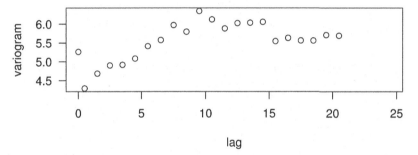

Fig. 8. Estimated variograms, assuming isotropy, for the temperatures and residual series respectively. The individual date series are treated as replicates.

Assessing temporal independence of time series residuals is a classic time series problem. Spectrum analysis is one way to address this issue. Fig. 7 shows the periodograms of the 81 individual residual time series and of their averaged series. Generally there appears to be some slopping down as one moves from the left to right. The AR(1) term was meant to handle this. This occurrence motivates the estimation of a multiplier

The bottom curve of Fig. 7 shows the average of the periodograms in the top panel. The horizontal line is as reference level. The plot is on log paper

There has to be some consideration of spatial dependence. The variogram is a classic parameter to study. An isotropic form will be estimated here. It may be considered an average of a non-isotropic form. Fig. 8 provides the results obtained via the R-function EmpiricalVariogram of the R-library RandomFields. This particular function accepts replicates. There are two panels in the figure, one for the residual series and the other for the temperature series. The y-axis scales in the two are quite different. This goes along with the inclusion of a monthly term in the model reduces the variability greatly.

Unfortunately no indication of uncertainty is provided. One might use some form of simulation or seek a multiplier, as in the temporal case. Fig. 8 suggests that spatial dependence extends some distance, e.g. up to lag 12. There was no use of universal transverse Mercator (see Wikipedia for a definition) coordinates, or of some other coordinate system in terms of km, so the lag units are confused. With the isotropic condition the correlation surface is ellipsoidally shaped. There is also the outlier at lag 0. It could be an artifact of the assumption of isotropy, or result from a mixture of populations being present. Lastly, one of the referees emphasized that the seasonal shape would be changing as one moved from the south to the north of the province. This effect has been dealt with in a sense by working with the individual station series.

6. Discussion/Conclusions

This paper has presented an approach for analyzing surface-air temperatures. Various limitations have turned up during it development. It is intended to address these in future work.

Time series development has often involved taking a statistic developed for the independent case and seeing its properties under a time series model. One can point to the work of Anderson,[1] Grenander and Rosenblatt,[7] Hannan,[8] with mean levels given by regression models and their study of the OLS estimate in the presence of stationary errors. Spatial-temporal dependence does appear to exist. If desired Gauss-Markov methods could be invoked to increase the efficiency of the estimates and to obtain more appropriate standard error values.

In practice the approach to studying spatial-temporal dependency has been to employ a model including some at the mean level and then check residuals for dependency hoping they turn out to be close to white noise. Here AR(1) modelling was introduced at the outset, following the result in Wood[14]. Yet temporal dependency beyond that appeared to exist. In consequence a "correction factor" was introduced. In the spatial case weights were introduced to attempt to reduce the spatial dependency, but an estimated variogram suggested that such dependency remained.

There are limitations of the present analysis. A key one is the non-general use of robustness methods. However a small early study of introducing biweights was carried out and the resulting figures did not change a lot. It needs also to be noted that temperature series from neighboring provinces/states were not included in the analyses and logically should be. Other explanatories, such as topography and station elevation might lead to improvements of the estimates, so too might interaction terms. One can mention month with latitude,

Other substantial contributions to spatial-temporal data modelling and analysis appear in: Cressie and Holan,[4] Dutilleul,[5] Gelfand *et al.*,[6] and Le and Zidek,[10] Jones and Vecchia,[9] and particularly Craigmile and Guttorp[3].

In conclusion, part of the intention of preparing this paper was to alert the statistical community to the presence of Berkeley Earth. Its data and the methods they are developing are going to provide a goldmine for statistical researchers. Members of that team have begun to exploit the data and preprints on the website, www.berkeleyearth.org. This paper shows that some "shovel ready" statistical software is available to employ. Having said that please note that this paper is not to be seen as a Berkeley Earth report.

Acknowledgments

I thank the Berkeley Earth Team, in particular R. Muller for inviting me to be involved, R. Rhode for developing and providing the data set employed this paper and C. Wickham for setting me up to use the statistical package R for some initial analyses.

Others to thank include M. Rooney who provided me with a listing of an R-program he had developed for spatial temporal data, Roger Byre of the Berkeley Department of Geography, R. Lovett, P. Spector, and R. Spence of the Berkeley Department of Statistics who assisted with the data management. The referees' gentle but firm comments altered the analyses performed substantially.

References

1. T. W. Anderson, *The Statistical Analysis of Time Series* (Wiley, New York, 1958).
2. Berkeley Earth, www.berkeleyearth.org (2012).
3. P. F. Craigmile and P. Guttorp, *J. Time Ser. Anal.* **32**, 378 (2011).
4. N. Cressie and S. H. Holan, *J. Time Ser. Anal.* **32**, 337 (2011).
5. P. Dutilleul, *Spatio-Temporal Heterogeneity: Concepts and Analyses* (Cambridge University Press, Cambridge, 2011).

6. A. E. Gelfand, P. J. Diggle, M. Fuentes and P. Guttorp, *Handbook of Spatial Statistics* (CRC Press, Boca Raton, 2010).

7. U. Grenander and M. Rosenblatt, *Statistical Analysis of Stationary Time Series* (Wiley, New York, 1957).

8. E. J. Hannan, *Time Series Analysis* (Methuen, London, 1960).

9. R. H. Jones and A. D. Vecchia, *J. Amer. Stat. Assoc.* **88**, 947 (1993).

10. N. D. Le and J. V. Zidek, *Statistical Analysis of Environmental Space-Time Processes* (Springer, New York, 2006).

11. R Development Core Team, *R: A Language and Environment for Statistical Computing,* http://www.R-project.org (R Foundation for Statistical Computing, Vienna, Austria, 2012).

12. R. Rohde, J. Curry, D. Groom, R. Jacobsen, R. A. Muller, S. Perlmutter, A. Rosenfeld, C. Wickham and J. Wurtele, http://berkeleyearth.org/pdf/berkeley-earth-santa-fe-robert-rohde.pdf (2011).

13. R. Rohde, J. Curry, D. Groom, R. Jacobsen, R. A. Muller, S. Perlmutter, A. Rosenfeld, C. Wickham and J. Wurtele. *Berkeley Earth temperature averaging process.* http://berkeleyearth.org/pdf/berkeley-earth-averaging-process.pdf (2011).

14. S. N. Wood, *Generalized Additive Models: An Introduction with R* (Chapman and Hall, London, 2006).

A MARKOV CHAIN MONTE CARLO SAMPLER FOR GENE GENEALOGIES CONDITIONAL ON HAPLOTYPE DATA

K. M. BURKETT, B. McNENEY and J. GRAHAM*

Department of Statistics and Actuarial Science, Simon Fraser University, 8888 University Drive, Burnaby, BC, V5A 1S6, Canada
* E-mail: jgraham@sfu.ca

The gene genealogy is a tree describing the ancestral relationships among genes sampled from unrelated individuals. Knowledge of the tree is useful for inference of population-genetic parameters such as migration or recombination rates. It also has potential application in gene-mapping, as individuals with similar trait values will tend to be more closely related genetically at the location of a trait-influencing mutation. One way to incorporate genealogical trees in genetic applications is to sample them conditional on observed genetic data. We have implemented a Markov chain Monte Carlo based genealogy sampler that conditions on observed haplotype data. Our implementation is based on an algorithm sketched by Zöllner and Pritchard[11] but with several differences described herein. We also provide insights from our interpretation of their description that were necessary for efficient implementation. Our sampler can be used to summarize the distribution of tree-based association statistics, such as case-clustering measures.

1. Introduction

The variation observed in the human genome is a result of stochastic evolutionary processes such as mutation and recombination acting over time. The gene genealogy for a sample of copies of a locus from unrelated individuals is a tree describing these ancestral events and relationships connecting the copies. Knowledge of the tree is useful for inference of population-genetic parameters and it also has potential application in gene-mapping. However, the time scale for genealogical trees is on the order of tens of thousands of years, and there is therefore no way to know the true underlying tree for a random sample of genes from a population.

One way to incorporate genealogical trees in genetic applications is to sample them conditional on observed genetic data, for example using

Markov chain Monte Carlo (MCMC) techniques. In population genetics, MCMC-based genealogy samplers have been implemented in order to estimate, for example, effective population sizes, migration rates and recombination rates. These approaches are reviewed in Stephens[10] and some software implementations are reviewed in Kuhner.[7] MCMC techniques have also been used in gene mapping methodology. In particular, Zöllner and Pritchard[11] implemented a coalescent-based mapping method in a program called LATAG (Local Approximation To the Ancestral Recombination Graph) that uses MCMC to sample genealogical trees. The LATAG approach involves sampling ancestral trees at a single focal point within a genomic region, rather than sampling the full ancestral recombination graph (ARG) (see Griffiths and Marjoram[3,4]) representation of the ancestral tree at *all* loci across the region. Focusing on a lower-dimensional latent variable, the genealogical tree, enables the LATAG approach to handle larger data sets than those that sample the ARG (see Zölner and Pritchard[11]).

In this paper, we describe our implementation of a genealogy sampler to sample ancestral trees at a locus conditional on observed haplotype data for surrounding SNPs. Since commonly-used genotyping technology does not provide haplotype information, haplotypes will typically be imputed from SNP genotype data. Our implementation is based on the genealogy sampler outlined by Zöllner and Pritchard,[11] herein called the ZP algorithm. As we were not interested in mapping *per se*, but rather in developing a stand-alone haplotype-based genealogy sampler that we could later extend to handle genotype data, we used their algorithm as a guide for developing our own haplotype-based sampler. During implementation, we filled in some of the details omitted from Zöllner and Pritchard[11] and made some modifications to the algorithm. We therefore provide a brief background on the ZP algorithm, describe some of the missing details, and highlight where our implementation differs. Although not described here, our sampler can be used to summarize the distribution of tree-based association statistics, such as case-clustering measures.

2. Notation

The observed data, $\mathbf{H}$, is a vector of size n consisting of a sample of haplotypes from unrelated individuals. Each haplotype, $\mathbf{s}_i$, is a sequence composed of the alleles at L SNP markers. The allele at the jth marker locus on the ith haplotype is denoted $s_{i,j}$ and can take one of two values, 0 or 1.

Due to recombination, the genealogy across a genomic region is represented by the ARG. However, the genealogy of a single locus is represented

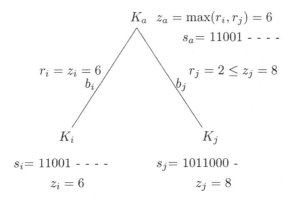

Fig. 1. Illustration of the definitions of the augmented variables on three nodes of τ_x. Each sequence is composed of $z_i - 1$ markers with alleles labeled 0 or 1. The focal point, x, is assumed to be to the left of the markers. Markers after recombination break points are labelled '-' and are not tracked because they are not co-inherited with the focal point.

by a tree. Rather than sample from the ARG, the ZP algorithm and our implementation both approximate the ARG by sampling marginal trees at single focal points, x. The genealogy then consists of a tree topology τ_x and the node times $\mathbf{T}$. The nodes of the tree are labelled K_i and the internal branches of the tree have length b_i. There are n tip nodes, one for each sampled haplotype, $\mathbf{s}_i$. Each of the $n - 1$ internal nodes corresponds to a coalescence event where two branches merge at a common ancestor. The root of the tree is the most recent common ancestor (MRCA) of the haplotypes in the sample. We are interested in sampling genealogies $(\tau_x, \mathbf{T})$ from the distribution $f(\tau_x, \mathbf{T}|\mathbf{H})$.

3. Overview of The ZP Algorithm

We now give a brief overview of the ZP algorithm in order to better compare their algorithm with our approach in later sections. We first provide details of the target distribution, $f(\tau_x, \mathbf{T}|\mathbf{H})$, for the genealogies of the focal point x conditional on the haplotype data. We then describe the MCMC approach to sample from this distribution. For more information about the ZP algorithm and the corresponding gene-mapping approach (see Zöllner and Pritchard[11]).

3.1. *A distribution for* $(\tau_x, \mathbf{T})$ *conditional on haplotypes* $\mathbf{H}$

In the ZP algorithm, the distribution for $(\tau_x, \mathbf{T})$, conditional on $\mathbf{H}$ is modeled by augmenting $(\tau_x, \mathbf{T})$ with parameters for the mutation rate θ and the recombination rate ρ, and latent variables for the unobserved sequence states at nodes of the tree. These unobserved states consist of the haplotypes at the internal node, $\mathbf{S} = (s_{n+1}, s_{n+2}, \ldots, s_{2n-1})$. However, only the sequence that is passed to the present along with the focal point x is stored, which requires also storing the location of the closest recombination events to the focal point x on either side of x. The recombination processes on either side of x are assumed to be independent, which simplifies the description since only variables and the model corresponding to the right-hand process need to be described. The variables and model for recombination events to the left of the focal point are all defined similarly to the right-hand versions. The recombination-related variables corresponding to the right-hand recombination process are denoted $\mathbf{R} = (r_1, r_2, \ldots, r_n, r_{n+1}, \ldots, r_{2n-1})$. More information about $\mathbf{R}$ is given below and an illustration of the definitions of $\mathbf{s}$ and r for the individual nodes is given in Fig. 1. Let $\mathbf{A} = (\tau_x, \mathbf{T}, \theta, \rho, \mathbf{S}, \mathbf{R})$.

With the addition of the latent variables, samples from $\mathbf{A}$ are now desired from target distribution $f(\mathbf{A}|\mathbf{H})$. This distribution is given by

$$f(\mathbf{A}|\mathbf{H}) \propto \Pr(\mathbf{S}|\mathbf{R}, \mathbf{T}, \tau_x, \theta, \gamma) \Pr(\mathbf{R}|\mathbf{T}, \tau_x, \rho) h(\tau_x, \mathbf{T}) h(\theta) h(\rho). \qquad (1)$$

The prior distribution for the topology and node times, $h(\tau_x, \mathbf{T})$, is the standard neutral coalescent model.[5,6] The prior distributions for the mutation and recombination rates, $h(\theta)$ and $h(\rho)$, are assumed to be uniform. The terms $\Pr(\mathbf{R}|\mathbf{T}, \tau_x, \rho)$ and $\Pr(\mathbf{S}|\mathbf{R}, \mathbf{T}, \tau_x, \theta, \gamma)$ are given below.

At each node, the sequence at a locus is stored only if that locus is co-inherited with the focal point. The sequence at the locus and the focal point will not be co-inherited when there is a recombination event between them. For node K_i, consider the maximum extent of sequence it leaves in at least one present descendant, and let z_i denote the index of the locus just beyond. Next, consider the maximum extent of sequence *from K_i's parent node* that K_i leaves in at least one present descendant, and let r_i be the index of the locus just beyond. In Fig. 1, nodes K_i and K_j are siblings with parent node K_a. From the definitions of z and r it follows that $z_a = \max(r_i, r_j)$. Moreover, r_a depends on the r values of all K_a's descendants only through the r values of K_a's two children or, more specifically, their maximum z_a.

By noting that the r value for a node depends on the r values of all its descendants through z, $\Pr(\mathbf{R}|\mathbf{T}, \tau_x, \rho)$ can be written as:

$$\Pr(\mathbf{R}|\mathbf{T}, \tau_x, \rho) = \prod_{i=1}^{2n-2} \Pr(r_i \mid z_i, b_i, \rho). \tag{2}$$

Recombination events along the branches of the tree and across the L loci are assumed to have a Poisson distribution with rate $\rho/2$ per unit of coalescence time. For each node then,

$$\Pr(r_i = c|z_i, b_i, \rho) = \begin{cases} 0 & c > z_i \\ \int_{d_{c-1}}^{d_c} \frac{b_i \rho}{2} \exp(-\frac{b_i \rho}{2} t) dt & 1 < c < z_i \\ \int_{d_{c-1}}^{\infty} \frac{b_i \rho}{2} \exp(-\frac{b_i \rho}{2} t) dt & c = z_i \end{cases}, \tag{3}$$

where the vector d consists of the locations of the L SNP markers.

The term $\Pr(\mathbf{S}|\mathbf{R}, \mathbf{T}, \tau_x, \theta, \gamma)$ is written by successively conditioning on parents:

$$\Pr(\mathbf{S}|\mathbf{R}, \mathbf{T}, \tau_x, \theta, \gamma) = \prod_{i=1}^{2n-1} \Pr(s_i|s_a, r_i, z_i, b_i, \theta, \gamma), \tag{4}$$

where K_a generically denotes the parent of K_i. We only track sequence that is co-inherited with the focal point, which consists of the sequence before the z_i^{th} marker. The first $r_i - 1$ loci have been inherited from K_a and therefore these loci are modeled by conditioning on the parental alleles. If the parental allele at a locus is different from its offspring's allele at the same locus, a mutation event has occurred on the branch between them. Mutation events at each locus along the branches of the tree are assumed to be Poisson distributed with rate $\theta/2$ per unit of coalescence time. Given that a mutation has occurred, the type of the new allele is chosen randomly from the two allelic types. Therefore, the probability of the $j^{th} < r_i$ allele is given by

$$\Pr(s_{i,j} = a_1|s_{a,j} = a_2, b_i, \theta) = \begin{cases} \frac{1}{2}(1 - e^{-\theta b_i/2}) & \text{if } a_1 \neq a_2 \\ \frac{1}{2}(1 - e^{-\theta b_i/2}) + e^{-\theta b_i/2} & \text{if } a_1 = a_2 \end{cases}, \tag{5}$$

where a_1 and a_2 are the allelic types at the jth marker. Loci from r_i to $z_i - 1$ have recombined in from an unknown ancestor. For these markers, Zöllner and Pritchard[11] assumed a first-order Markov model where the allele at a locus has a Bernoulli distribution with probability that depends on the allele at the previous locus. These probabilities, denoted γ, are estimated from the observed data. Therefore, for the ith node, each term in Eqn. (4)

is

$$\Pr(\mathbf{s}_i|\mathbf{s}_a, r_i, z_i, b_i, \theta, \gamma) = \left[\prod_{j=1}^{r_i-1} \Pr(s_{i,j}|s_{a,j}, b_i, \theta)\right] \Pr(h_{r_i \to z_i - 1}|\gamma), \quad (6)$$

where $h_{r_i \to z_i - 1}$ denotes the haplotype of the sequence between the r_i^{th} and $z_i - 1^{th}$ locus and $\Pr(s_{i,j}|s_{a,j}, b_i, \theta)$ is given in Eqn. (5).

3.2. Overview of the proposal distributions used in the ZP algorithm

MCMC is used to sample from the target distribution, $f(\mathbf{A}|\mathbf{H})$. A new value for $\mathbf{A}$, $\tilde{\mathbf{A}}$, is proposed from distribution $Q(\tilde{\mathbf{A}}|\mathbf{A})$. This value is then accepted or rejected according to the Metropolis-Hastings acceptance probability

$$\alpha(\mathbf{A}, \tilde{\mathbf{A}}) = \min\left\{1, \frac{f(\tilde{\mathbf{A}}|\mathbf{H})Q(\mathbf{A}|\tilde{\mathbf{A}})}{f(\mathbf{A}|\mathbf{H})Q(\tilde{\mathbf{A}}|\mathbf{A})}\right\}. \quad (7)$$

The ZP algorithm uses six update schemes to propose new values for $\mathbf{A}$. Each of the six schemes proposes new values for only a subset of the components of $\mathbf{A}$. The six update schemes are: (1) Update θ; (2) update ρ; (3) local update of internal nodes; (4) major topology change; (5) minor topology change and (6) reorder coalescence events. These update schemes are illustrated in Fig. 2 and are summarized below.

(1) **Update θ:** Sample $\tilde{\theta}$ from a uniform distribution on $(\theta^{(t)}/2, 2\theta^{(t)})$, where $\theta^{(t)}$ is the value of θ at the tth iteration.

(2) **Update ρ:** Sample $\tilde{\rho}$ from a uniform distribution on $(\rho^{(t)}/2, 2\rho^{(t)})$, where $\rho^{(t)}$ is the value of ρ at the tth iteration.

(3) **Local updates of internal nodes:** For each node, starting at the tip nodes and ending at the MRCA, sample a new time since the present, $\tilde{t}_i$, recombination variable, $\tilde{r}_i$, and new sequence, $\tilde{s}_i$. The proposal distribution for sampling each component was not provided.

(4) **Major topology change:** Randomly select a node to be moved, K_{c_1} in Fig. 2(A), from the set of nodes that can be moved without causing incompatible r values among neighbouring nodes. Referring to Fig. 2(C), after the topology change, loci 1 through $z_p - 1$ must pass to the present via either K_{c_2} or K_a. An incompatibility can occur if $z_p > \max(r_a, z_{c_2})$. A new sibling, K_s, is selected from the set of nodes having parents older than t_{c_1} and based on sequence similarity to s_{c_1}. The topology change is made, as illustrated in Fig. 2(C), a new time

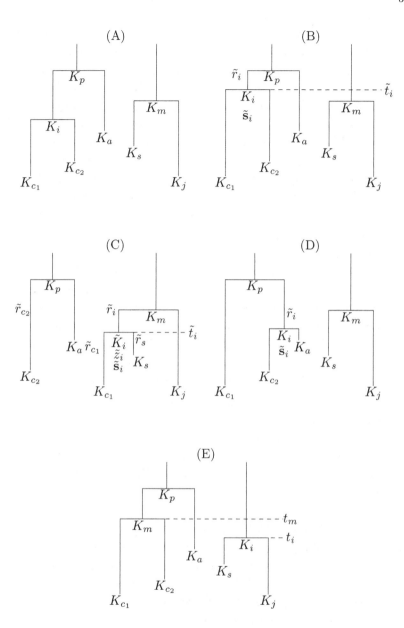

Fig. 2. Illustration of the update schemes for the ZP algorithm. (A) Example of two branches of the tree before any updates; (B) After local update to node K_i; (C) After a major topology change to node K_{c_1}; (D) After a minor topology change to node K_{c_1}; (E) After reordering coalescence events for nodes K_i and K_m. To highlight the nodes and branches modified during each update, those that remain the same are greyed out.

for the parent $\tilde{K}_i$ is sampled from a uniform distribution with bounds $(\max(t_{c_1}, t_s), t_m)$, new r values are sampled for nodes K_{c_1}, K_{c_2}, K_s and $\tilde{K}_i$, and a new sequence is sampled for $\tilde{K}_i$.

(5) **Minor topology change:** An internal node, K_{c_1} in Fig. 2(A), is first sampled. In the new topology its old sibling, K_{c_2}, and old aunt, K_a, become its new nieces. After the topology change, illustrated in Fig. 2(D), r and **s** value are sampled for the new parent node $\tilde{K}_i$.

(6) **Reorder coalescence events:** The order of coalescence events is modified by swapping the times for two nodes, K_i and K_m, as shown in Fig. 2(E). The two nodes are eligible to swap node times provided that, after the swap of times, K_i and K_j are not older than their respective parents or younger than their respective children.

At each iteration of the chain, combinations of these update schemes are used to propose new values for $\mathbf{A}$; for example, at the tth iteration a major topology change and local updates might be used to propose $\tilde{\mathbf{A}}$. Information about the frequency with which each combination of updates is selected is not given, but it is stated that the ZP algorithm samples each combination with a probability that depends on the nature of the dataset.

4. Description of Our Genealogy Sampler

In this section, we describe our haplotype-based genealogy sampler. As mentioned, we based our approach on that of Zöllner and Pritchard[11] and therefore many of the details given in Sec. 3 apply equally to our sampler. In the description that follows, we therefore highlight the differences between our approach and that of the ZP algorithm. A more comprehensive description of our sampler can also be found in Ref.[1]

4.1. *The target distribution:* $f(\mathbf{A}|\mathbf{H})$

The target distribution is modeled as in Sec. 3.1, but we use an alternate prior distribution for the recombination rate, ρ. We do not assume a uniform prior distribution for ρ as there is some evidence that the distribution of recombination rates is roughly exponential in parts of the genome.[8] We therefore assume a gamma/exponential prior distribution for ρ.

4.2. *Proposal distributions for* t_i, r_i *and* $\mathbf{s}_i$

The local updates and the topology rearrangements sample new t, r, and **s** values for some nodes of the tree; however, the proposal distributions for sampling these new values were not given in Zölner and Pritchard.[11] We therefore now provide information on our proposal distributions for these components. For a node K_i, our proposal distributions for t_i, r_i and $\mathbf{s}_i$ are all motivated by the decomposition of $f(\mathbf{A}|\mathbf{H})$ given in Eqn. (1). That is, t_i is sampled conditional on other t values from an approximation to $h(t_i|T_{-i})$, r_i from an approximation to $\Pr(r_i|\mathbf{R}_{-i}, \mathbf{T}, \tau_x, \rho)$ and $\mathbf{s}_i$ from $\Pr(\mathbf{s}_i|\mathbf{S}_{-i}, \mathbf{R}, \mathbf{T}, \tau_x, \theta, \gamma)$.

Proposal distribution for t_i

The proposal distribution for t_i is motivated by the distribution $h(t_i|\mathbf{T}_{-i})$; however, rather than condition on all elements of $\mathbf{T}_{-i}$ in the proposal distribution, we condition only on the times of the nodes adjacent to K_i. Letting K_{c_1} and K_{c_2} be K_i's two children, and K_a be K_i's parent (see Fig. 2(A) for the node labels), the proposal distribution for t_i is

$$q(t_i|t_{c_1}, t_{c_2}, t_a) = \begin{cases} \exp(-(t_i - t_{2n-2}))I[t_i > t_{2n-2}] & i = 2n - 1 \\ U(\max(t_{c_1}, t_{c_2}), t_a) & i \neq 2n - 1 \end{cases}. \quad (8)$$

The motivation for this proposal distribution comes from the coalescent model, which assumes exponential inter-coalescence times with rate $\binom{j}{2}$ when there are j lineages left to coalesce. If K_i is the MRCA, *i.e.* $i = 2n - 1$, $j = 2$ and the waiting time until the last two lineages merge, $t_i - t_{i-1} = t_{2n-1} - t_{2n-2}$, has an $\exp(1)$ distribution. On the other hand, if K_i is not the MRCA, the distribution of t_i conditional on t_{i-1} and t_{i+1} is uniform on (t_{i-1}, t_{i+1}). Rather than condition on the times of the adjacent coalescence events, *i.e.* t_{i-1} and t_{i+1}, we condition on the times of the adjacent nodes, t_{c_1}, t_{c_2} and t_a. Therefore, our uniform proposal distribution only approximates the true conditional distribution. However, this proposal distribution does allow for a re-ordering of coalescence events since the proposal distribution for t_i is not necessarily constrained by t_{i-1} and t_{i+1}.

Proposal distribution for r_i

Recall that the r variables are defined recursively so that the value of r_i depends on the r values of all of its descendants through z_i. In the description of the model for the r values given in Zöllner and Pritchard [11] (page 1090), they explicitly let the probability of r_i, given in Eqn. (3) here, be 0 if the value is incompatible with r values of ancestral nodes. Therefore, the ZP algorithm seems to allow an $\tilde{r}_i$ to be sampled that is incompatible with other r values on the tree; such $\tilde{r}_i$ values would be immediately rejected as they are defined to have 0 probability under the target distribution.

In our updates to the r_i variables, rather than propose values that would be rejected due to incompatibility, we restrict the support of the proposed

r_i to be compatible with all other r variables on the tree. The proposed value for r_i is constrained by the length of sequence that is passed to the present through node K_i; that is, by z_i. The proposed value may also be constrained by r_a, the r value of an ancestor K_a, if the sequence material passes to the present through K_i. Therefore, the support for the proposal distribution for r_i, $\mathcal{S}(\mathbf{R}_{-i})$, depends on restrictions imposed by the r and z values of nodes in the vicinity of node K_i. The actual nodes that restrict the support depend on the update type and therefore more information about the restrictions is provided in Sec. 4.3.

With the support $\mathcal{S}(\mathbf{R}_{-i})$ giving the set of values for r_i that are compatible with the data at surrounding nodes, the conditional distribution for component r_i is

$$\Pr(r_i|\mathcal{S}(\mathbf{R}_{-i}), \mathbf{T}, \tau_x, \rho) = \frac{\Pr(r_i|z_i, b_i, \rho)}{\sum_{r^* \in \mathcal{S}(\mathbf{R}_i)} \Pr(r^*|z_i, b_i, \rho)} 1[r_i \in \mathcal{S}_i(\mathbf{R}_{-i})].$$

We thus take as our proposal distribution for r_i

$$q(r_i|\mathcal{S}(\mathbf{R}_{-i}), \mathbf{T}, \tau_x, \rho) = \Pr(r_i|\mathcal{S}(\mathbf{R}_{-i}), \mathbf{T}, \tau_x, \rho), \tag{9}$$

where $\Pr(r_i|z_i, b_i, \rho)$ is given in Eqn. (3).

Proposal distributions for $\mathbf{s}_i$

For our proposal distribution for component $\mathbf{s}_i$, recall that $\mathbf{s}_i$ is a vector, with L elements corresponding to the alleles at each locus. We therefore sample a new sequence, $\tilde{\mathbf{s}}_i$, by sampling a new allele at each locus $s_{i,j}$ starting at the first locus, $j = 1$, and finishing at $j = z_i - 1$. Since we are only sampling alleles at loci that are passed to at least one descendant at present in the tree, all alleles at markers z_i and above are given the value '-' with probability one, as shown in Fig. 1.

To motivate a proposal distribution for $s_{i,j}$, the allele at the jth locus, we use information available from the allele at the $j - 1^{th}$ locus, $s_{i,j-1}$, and the alleles at the jth locus in the sequences of adjacent nodes. If the jth locus was inherited from parent K_a, i.e. if $j < r_i$, then $s_{i,j}$ will be different from $s_{a,j}$ if a mutation occurred on the branch b_i. The probability of mutation events on branches of the tree was given in Eqn. (5). If this locus was not inherited from K_a, i.e. if $j \geq r_i$, then, by the first-order Markov haplotype model referred to in Eqn. (6), the probability of the allele depends on the allele at the $j - 1^{th}$ locus. Similar arguments can be made for the inheritance of this locus between K_i and its children, K_{c_1} and K_{c_2}.

Our proposal distribution for $s_{i,j}$ is motivated by combining the inheritance of the jth locus from K_a through nodes K_i, K_{c_1} and K_{c_2}. Let the proposal distribution for the jth locus be

$$q(s_{i,j}|s_{i,j-1}, s_{c_1,j}, s_{c_2,j}, s_{a,j}, \mathbf{R}, \mathbf{T}, \tau_x, \theta, \gamma)$$
$$= \Pr(s_{i,j}|s_{i,j-1}, s_{c_1,j}, s_{c_2,j}, s_{a,j}, \mathbf{R}, \mathbf{T}, \tau_x, \theta, \gamma)$$
$$\propto \Pr(s_{i,j}, s_{c_1,j}, s_{c_2,j}|s_{i,j-1}, s_{a,j}, \mathbf{R}, \mathbf{T}, \tau_x, \theta, \gamma)$$
$$= \Pr(s_{i,j}|s_{i,j-1}, s_{a,j}, \mathbf{R}, \mathbf{T}, \tau_x, \theta, \gamma) \times \Pr(s_{c_1,j}|s_{i,j}, \mathbf{R}, \mathbf{T}, \tau_x, \theta, \gamma) \times$$
$$\Pr(s_{c_2,j}|s_{i,j}, \mathbf{R}, \mathbf{T}, \tau_x, \theta, \gamma). \tag{10}$$

The last line follows from the conditional independence of $s_{c_1,j}$ and $s_{c_2,j}$ given $s_{i,j}$, their parent's allele at the jth locus.

If K_i is not the MRCA ($i \neq 2n - 1$), the first term in Eqn. (10) is

$$\Pr(s_{i,j}|s_{i,j-1}, s_{a,j}, \mathbf{R}, \mathbf{T}, \theta, \gamma) = \begin{cases} \Pr(s_{i,j}|s_{a,j}, b_i, \theta) & j < r_i \\ \Pr(s_{i,j}|\gamma) & j = r_i \\ \Pr(s_{i,j}|s_{i,j-1}, \gamma) & r_i < j < z_i - 1 \end{cases},$$

with $\Pr(s_{i,j}|s_{a,j}, b_i, \theta)$ given in Eqn. (5), and $\Pr(s_{i,j}|\gamma)$ and $\Pr(s_{i,j}|s_{i,j-1}, \gamma)$ given by the first-order Markov haplotype model referred to in Eqn. (6). If K_i is the MRCA, then K_i does not have a parent in the tree and therefore the probability of the jth locus is based on the haplotype model:

$$\Pr(s_{i,j}|s_{i,j-1}, \mathbf{R}, \mathbf{T}, \theta, \gamma) = \begin{cases} \Pr(s_{2n-1,j}|\gamma) & j = 1 \\ \Pr(s_{2n-1,j}|s_{2n-1,j-1}\gamma) & 1 < j \leq z_i - 1 \end{cases}.$$

The second and third terms of Eqn. (10) are modeled similarly. However, if K_i does not pass the jth locus to K_{c_1} and/or K_{c_2} then these nodes provide no information about $s_{i,j}$. Hence the probabilities $\Pr(s_{c_1,j}|s_{i,j}, \mathbf{R}, \mathbf{T}, \tau_x, \theta)$ and $\Pr(s_{c_2,j}|s_{i,j}, \mathbf{R}, \mathbf{T}, \tau_x, \theta)$ are constant with respect to $s_{i,j}$ and so can be dropped. It follows that the second and third terms are:

$$\Pr(s_{c_1,j}|s_{i,j}, \mathbf{R}, \mathbf{T}, \theta) = \begin{cases} \Pr(s_{c_1,j}|s_{i,j}, b_{c_1}, \theta) & j < r_{c_1} \\ 1 & j \geq r_{c_1} \end{cases} \text{ and}$$

$$\Pr(s_{c_2,j}|s_{i,j}, \mathbf{R}, \mathbf{T}, \theta) = \begin{cases} \Pr(s_{c_2,j}|s_{i,j}, b_{c_2}, \theta) & j < r_{c_2} \\ 1 & j \geq r_{c_2} \end{cases}.$$

To summarize, the proposal distribution for $s_{i,j}$, given in Eqn. (10), is proportional to the product of the probabilities for the inheritance of the jth locus from its parent K_a to its children K_{c_1} and K_{c_2}. The proposal probability for the full sequence, $\tilde{s}_i$, is proportional to the product of the proposal probabilities for each allele at each locus.

4.3. Update schemes for our sampler

As mentioned, we used the outline given for the six update schemes, which are summarized in Sec. 3.2, as a starting point for our sampler. During implementation, we made some changes to the updates schemes. First, we generally aimed to increase sampling efficiency by not proposing values that would result in the update being automatically rejected due to incompatibility with the unmodified values. For example, referring to Fig. 2(A), this can occur if $\tilde{r}_i$ is lower than the r value for K_i's parent K_a. An incompatibility can also occur with the sequences, $\mathbf{s}$, due to an update to the r's. For example, if $\tilde{r}_i$ leads to $\tilde{z}_a > z_a$ and $\mathbf{s}_a$ is not also updated, then the current value for $\mathbf{s}_a$ will not be possible since it will not have valid alleles at markers z_a to $\tilde{z}_a - 1$. Second, our algorithm does not use the sixth update, which reorders coalescence events by swapping node times, since the proposal distribution we use to update the t_i allows for a reordering of coalescence events. Finally, by default, at each iteration of the chain, only one of the five updates is used to propose new $\mathbf{A}$ rather than a combinations of the six; however, in our implementation the user can choose to perform combination of updates consecutively. Our modifications to the update schemes are now described.

(1) **Updating θ:** Our proposal distribution for θ is also uniform. However, the range of the uniform is set so that θ outside of the prior distribution, which would be rejected, can not be sampled.

(2) **Local updates:** In the local updates of the ZP algorithm, new values for t_i, r_i and $\mathbf{s}_i$ are proposed for each node, starting at the tip nodes and moving to the MRCA. Our local updates proceed from the first internal node to the MRCA and we choose to accept or reject the changes at each node before moving to the next node. At each internal node, K_i, we propose a new time, t_i, from the proposal distribution given in Sec. 4.2. We then propose new r_{c_1} and r_{c_2}, for the children K_{c_1} and K_{c_2} rather than for K_i, from proposal distribution given in the Sec. 4.2. We restrict the support of $\tilde{r}_{c_1}$ and $\tilde{r}_{c_2}$ so that the value of r_i is not incompatible with $\tilde{r}_{c_1}$ and $\tilde{r}_{c_2}$ (*i.e.* one of $\tilde{r}_{c_1}$ and $\tilde{r}_{c_2}$ must be greater than or equal to r_i). Finally, we propose new $\mathbf{s}_i$ from the proposal distribution given in Sec. 4.2. Our modifications to the local updates ensure that (1) the sampled r values are not immediately rejected due to incompatibility with other r values and (2) an update to $\mathbf{s}_a$ is not required (if r_i were updated, then by definition z_a might also change).

(3) **Major topology change:** We made several changes to the major topology rearrangement:

(a) The similarity score used in the ZP algorithm was not provided. Our similarity score between two sequences, $\mathbf{s}_i$ and $\mathbf{s}_c$, is

$$p_{ic} = \begin{cases} \dfrac{\sum_{j=1}^{\min(z_i-1,z_c-1)} 1(s_{i,j}=s_{c,j})}{\min(z_i-1,z_c-1)} & K_c \text{ is eligible to coalesce with } K_i \\ 0 & \text{otherwise} \end{cases}.$$

This score counts the alleles that $\mathbf{s}_i$ and $\mathbf{s}_c$ share in common. Ideally, we would only like to compare loci that both sequences inherited from their shared parent. However, when selecting the new sibling, we don't have this information as $\tilde{r}_i$ and $\tilde{r}_c$ haven't yet been sampled. We therefore only compare the loci that both would have inherited from their parent if no recombination events occur on the branches $\tilde{b}_i$ and $\tilde{b}_c$. This sequence is between locus 1 and $\min(z_i-1, z_c-1)$.

(b) We also exclude nodes in selecting K_s in order to avoid creating non-sensical trees and to differentiate the major from the minor topology change; the excluded nodes are the current sibling, parent, aunt, niece or grandparent of K_{c_1}.

(c) As with the local updates, we have structured our updates to the r variables so that the proposed values do not cause incompatibilities with unchanged values on the tree. Referring to Fig. 2(A) and (C) for the notation, this is done by restricting the support of $\tilde{r}_{c_2}$ so that $\tilde{z}_p = z_p$ and the supports of $\tilde{r}_s$ and $\tilde{r}_i$ so that $\tilde{z}_m = z_m$.

(4) **Minor topology change:** Our implementation of the minor topology change also differs substantially:

(a) Referring to Fig. 2(A) and (D), we impose the condition that $t_i > t_a$, or K_i must be older than its sibling K_a, to ensure that the topology change produces a valid tree. Since t_i and t_a are the same after the topology change, without this condition the parent $\tilde{K}_i$ could be younger than its child K_a and the update would be automatically rejected.

(b) The ZP algorithm imposed a constraint that K_{c_1} be an internal node. However, if tip nodes are not eligible for the topology change, the reverse rearrangement of $\tilde{\mathbf{A}}$ to $\mathbf{A}$ could have zero probability, the acceptance probability would be 0, and the update would automatically be rejected. We therefore remove this restriction.

(c) Even though the time of node $\tilde{K}_i$ is the same in $\tilde{\tau}_x$ as that of K_i in τ_x, the branch lengths b_{c_1} and b_a are not the same. For this reason, r_{c_1} and r_a could be unlikely given the new branch lengths. We therefore also sample new r_{c_1} and r_a.

(d) As with the major topology change and the local updates, we sample r values in such a way that compatibility with other r values is ensured. For this update, this is achieved if we require that $\tilde{z}_p = z_p$. This results in a restricted support so that one of $\ddot{r}_{c_1}$ or $\ddot{r}_a$ is greater than or equal to z_p.

4.4. *Software and application*

The sampler was coded in C++ and currently runs at the command line; however, we plan to import the C++ code into R. Input options are provided to the program in a file and include the file names for the relevant datasets, a run name, chain length, burn-in, thinning, as well as prior parameters and initial values for the latent variables. Most options include sensible defaults. Output includes the scalar-valued output (the update performed, an acceptance indicator and the ρ and θ values) and the sampled trees.

The sampler was applied to haplotype data imputed from the Crohn's dataset[2] available in the R gap package. Each dataset consisted of 516 haplotypes of between 20 and 35 SNPs and was run on a separate 2.67 GHz core of a cluster computer. Eight million iterations, which took approximately three weeks, were required to achieve convergence. Due to space constraints, we refer the reader to Sec. 2.7 of Burkett[1] for the full results, including convergence diagnostics.

5. Discussion

In this paper, we have given a summary of our haplotype-based sampler. Our sampler is based on the approach outlined in Ref. 11 but we have provided some important details related to our implementation. In particular, all our proposal distributions for the recombination variable r lead to proposed values that do not cause incompatibilities with other variables on the tree and hence have non-zero probabilities under the target distribution. In contrast, the ZP algorithm can propose r values that are incompatible with ancestral nodes since they explicitly define the probability of such events to be zero under the target distribution. This would be inefficient as it would lead to the proposal being rejected. A related point is that we structured the local updates so that sequence at an ancestral node doesn't become incompatible with the proposed values. In further efforts to improve efficiency, we also modified the proposal distribution for the minor topology rearrangement so that transitions from $\tilde{A}$ back to A are always possible. We have also eliminated the need for a separate update to reorder coalescence events. Finally, by default, we do not perform sets of updates in a sequential order, for example a major topology rearrangement followed by the local updates; we have found that performance of the sampler is unaffected by this simplification.

Our sampler returns the tree and node times in Newick format. The trees can be read into existing software packages for graphical display and analysis. Any tree-based statistic that relies on the topology and node times, such as the case-clustering measure from Ref.,[9] can be computed from a sampled tree. The distribution of the statistic can be summarized by computing the statistic on the MCMC sample of trees. More information and an example of this approach is provided in Burkett.[1]

A drawback to using a genealogical sampler that conditions on haplotype data is that with SNP data the haplotypes are often unknown. Therefore, the haplotypes are also latent data in this context. The suggestion in Zöllner and Pritchard[11] was to first use a program to statistically impute phase. However, since we are sampling trees based on a single imputation of the haplotypes, the imputation limits the space of trees that we sample from which could subsequently bias estimates. We are therefore working on an extension to the sampler to handle missing phase.

Acknowledgments

This work was supported by the Natural Sciences and Engineering Research Council of Canada and the Mathematics of Information Technology and Complex Systems Networks of Centres of Excellence. Portions of this work were undertaken while KB held a Canadian Institutes of Health Research Doctoral Research Award and a Michael Smith Foundation for Health Research (MSFHR) Senior Graduate Trainee Award, and while JG was a MSFHR Scholar.

References

1. K. M. Burkett, *Markov Chain Monte Carlo Sampling of Gene Genealogies Conditional on Observed Genetic Data*, PhD Thesis, Simon Fraser University, (Burnaby, Canada, 2011).
2. M. J. Daly, J. D. Rioux, S. F. Schaffner, T. J. Hudson and E. S. Lander, *Nature Genetics* **29**, 229 (2001).
3. R. C. Griffiths and P. Marjoram, *Journal of Computational Biology* **3**, 479 (1996).
4. R. C. Griffiths and P. Marjoram, An Ancestral Recombination Graph, In *Progress in Population Genetics and Human Evolution*, Eds.: P. Donnelly and Tavare, (Springer-Verlag, New York, 1997) 257-270.
5. R. R. Hudson, *Theoretical Population Biology* **23**, 183 (1983).

6. J. F. C. Kingman, *Stochastic Processes and their Applications* **13**, 235 (1982).
7. M. K. Kuhner, *Trends in Ecology & Evolution* **24**, 86 (2009).
8. G. A. T. McVean, S. R. Myers, S. Hunt, P. Deloukas, D. R. Bentley and P. Donnelly, *Science* **304**, 581 (2004).
9. M. J. Minichiello and R. Durbin, *American Journal of Human Genetics* **79**, 910 (2006).
10. M. Stephens, Inference Under the Coalescent, In *Handbook of Statistical Genetics, Second Edition*, Eds.: D. J. Balding, M. Bishop and C. Cannings (John Wiley and Sons, London, 2003), 636-661.
11. S. Zöllner and J. K. Pritchard, *Genetics* **169**, 1071 (2005).

JACKKNIFING STOCHASTIC RESTRICTED RIDGE ESTIMATOR WITH HETEROSCEDASTIC ERRORS

YOGENDRA P. CHAUBEY

Department of Mathematics and Statistics, Concordia University,
Montréal, H3G 1M8, Canada
E-mail: chaubey@alcor.concordia.ca

MANSI KHURANA*

Department of Mathematics and Statistics, Banasthali University,
Rajasthan, 304022, India
E-mail: kmansi@banasthali.in

SHALINI CHANDRA

Indian Statistical Institute, North East Centre,
Tezpur, 784028, India
E-mail: cshalini@isine.ac.in

In the present article, we propose a Jackknifed version of stochastic restricted ridge regression (SRR) estimator given by Özkale[21] on the lines of Singh *et al.*[25]. The performance of Jackknifed estimator is investigated in terms of bias and mean square error with those of SRR estimator theoretically as well as with a numerical example. The results have also been validated using a Monte Carlo study.

1. Introduction

Consider the following multiple linear regression model

$$y = X\beta + u \tag{1.1}$$

where y is an $(n \times 1)$ vector of observations on the variable to be explained, X is an $(n \times p)$ matrix of n observations on p explanatory variables assumed to be of full column rank, β is a $(p \times 1)$ vector of regression coefficients associated with them and u is an $(n \times 1)$ vector of disturbances, the elements of

---

*This work was completed while the author was visiting Concordia University on a Common Wealth visiting scholarship.

which are assumed to be independently and identically normally distributed with

$$E(u) = 0; \quad Var(u) = \sigma^2 I. \tag{1.2}$$

The ordinary least squares (OLS) estimator, $\hat{\beta} = (X'X)^{-1}X'y$ is well known to be 'optimum' in the sense that $\ell'\hat{\beta}$ provides the minimum variance unbiased estimator of $\ell'\beta$ among the class of all linear unbiased estimators. However, in case of ill conditioned design matrix or multicollinearity, OLS estimator produces huge sampling variances which may result in the exclusion of significant variables from the model (see Farrar and Glauber[3]). In this situation, we switch to ridge regression (RR) estimator given by Hoerl and Kennard[8] which is of the following form

$$\hat{\beta}_R = (X'X + kI_p)^{-1}X'y; \quad k > 0. \tag{1.3}$$

It was demonstrated by Hoerl and Kennard[8] that there always exists a $k > 0$ such that $\hat{\beta}_R$ performs better than $\hat{\beta}$ in terms of mean square error, however, as pointed in Singh et $al.$[25] it may carry serious bias. Another solution to the problem of multicollinearity is the use of prior information available in the form of exact or stochastic restrictions binding the regression coefficients. If the restrictions on a particular parameter or a linear combination of parameters are exact and of the following form

$$R\beta = r \tag{1.4}$$

where r is an $(m \times 1)$ vector of known elements and R is an $(m \times p)$ known prior information on regression coefficients of rank $m \leq p$, the well-known restricted least squares (RLS) estimator is given by

$$\hat{\beta}_{RL} = \hat{\beta} - (X'X)^{-1}R'[R(X'X)^{-1}R']^{-1}(R\hat{\beta} - r). \tag{1.5}$$

This is the 'optimum' estimator of β in the class of linear and unbiased estimators, in the sense described earlier, under the restriction given by Eq. (1.4). Combining the ideas of both ridge regression and RLS estimators, Sarkar[24] proposed the restricted ridge regression (RRR) estimator and investigated its superiority over the other two estimators. However, it was pointed out by Groß[6] that RRR estimator does not satisfy the linear restriction in Eq. (1.4) for every outcome of y and therefore he introduced a new restricted ridge regression estimator.

There are many situations in applied field when the exact prior information on coefficient vector is quite often not appropriate and sometimes, this type of information may not even exist. In such situations, the set of constraints on coefficients may be treated as probability statements which may be converted in the form of stochastic linear constraints of the following form

$$r = R\beta + \phi \tag{1.6}$$

where r is an $(m \times 1)$ random vector of known elements and R is an $(m \times p)$ known prior information on regression coefficients of rank $m \leq p$ and ϕ is a random vector independent of u, with $E(\phi) = 0$ and $Var(\phi) = \sigma^2 W$, where W is an $m \times m$ known positive definite matrix.

From here onwards, we will not make use of the assumption in Eq. (1.2); instead, we are going to consider the more general case of heteroscedastic or autocorrelated errors. Thus we consider the model in Eq. (1.1) with

$$E(u) = 0; \quad Var(u) = \sigma^2 V \tag{1.7}$$

where V is an $n \times n$ known positive definite symmetric matrix. As V is positive definite, there exists a nonsingular symmetric matrix F such that $V^{-1} = F'F$. Now pre-multiplying the model in Eq. (1.1) with F and following new assumptions in Eq. (1.7), we get

$$Fy = FX\beta + Fu. \tag{1.8}$$

The OLS estimator of β for model in Eq. (1.8) will be of the following form

$$\hat{\beta}_{GL} = (X'V^{-1}X)^{-1}X'V^{-1}y \tag{1.9}$$

which is known as the generalized least squares (GLS) estimator of β for model in Eq. (1.1) with Eq. (1.7) (see Rao and Toutenburg[23]). This is unbiased and enjoys minimum variance among the class of linear unbiased estimators. For dealing with multicollinearity in this case when $X'V^{-1}X$ becomes ill conditioned and GLS produces large sampling variances, Trenkler[28] proposed the ridge estimator of β for model in Eq. (1.1) with assumptions in Eq. (1.7) as

$$\hat{\beta}_{GL}(k) = (X'V^{-1}X + kI_p)^{-1}X'V^{-1}y; \quad k > 0. \tag{1.10}$$

Now, we focus our attention on the stochastic restrictions in Eq. (1.6) that may help in solving the problem of multicollinearity. Keeping this in mind, Theil and Goldberger[27] proposed a new technique to estimate the parameters of model in Eq. (1.1) with restrictions in Eq. (1.6) which they called *mixed estimation*. This Bayes like technique is so termed as the prior or non sample information on some or all the parameters of the model is directly combined with the sample information. To obtain the mixed regression (MR) estimator, they augmented y and X as

$$\begin{pmatrix} y \\ r \end{pmatrix} = \begin{pmatrix} X \\ R \end{pmatrix} \beta + \begin{pmatrix} u \\ \phi \end{pmatrix} \tag{1.11}$$

or

$$y_m = X_m \beta + u_m \tag{1.12}$$

where $E(u_m) = 0$ and $Var(u_m) = \sigma^2 \tilde{W} = \sigma^2 \begin{pmatrix} V & 0 \\ 0' & W \end{pmatrix}$, $\tilde{W}$ is positive definite as V and W are also positive definite. Since $\tilde{W}$ is positive definite, there exists a nonsingular symmetric matrix N such that $\tilde{W} = N'N$. Premultiplying both sides of Eq. (1.12) by N^{-1}, we get

$$y_* = X_* \beta + u_* \tag{1.13}$$

where $y_* = N^{-1} y_m$, $X_* = N^{-1} X_m$, $u_* = N^{-1} u_m$ and $Var(u_*) = \sigma^2 I$. So, we find that Eq. (1.13) is a form of standard linear regression model.

Theil and Goldberger[27] minimized the sum of squared residuals in this model and obtained the MR estimator which has the following form

$$\hat{\beta}_m = (X'_* X_*)^{-1} X'_* y_* = (X'V^{-1}X + R'W^{-1}R)^{-1}(X'V^{-1}y + R'W^{-1}r). \tag{1.14}$$

They demonstrated that MR estimator performed better than the classical least squares estimator when sample error variance and the covariance matrix of the stochastic prior information is known. Nagar and Kakwani[17] extended their work and considered the case when prior information may be available in the form of lower and upper stochastic bounds on coefficients. In a recent paper, Özkale[21] modified the MR estimator on the lines of Trenkler[28] and proposed a new estimator namely SRR estimator which is given by

$$\hat{\beta}_k = (X'_* X_* + kI_p)^{-1} X'_* y_*$$
$$= (X'V^{-1}X + R'W^{-1}R + kI_p)^{-1}(X'V^{-1}y + R'W^{-1}r). \tag{1.15}$$

The conditions under which SRR estimator dominates MR estimator in terms of mean square error were obtained and the results were verified with a numerical example. However like ridge, SRR estimator also carries a substantial amount of bias which had been ignored in the study. In the next section, in order to control the bias, we define the Jackknifed form of SRR estimator following the lines of Singh et al.[25] and Singh and Chaubey[26] for the transformed model.

2. Jackknifed SRR Estimator

To induce simplicity, we are going to use the canonical form. Since $X'_* X_* = X'_m \tilde{W}^{-1} X_m = X'V^{-1}X + R'W^{-1}R$ is a $p \times p$ symmetric matrix, so there exists a $p \times p$ orthogonal matrix G such that $G'G = GG' = I$ and $G'X'_* X_* G = \Lambda = diag(\lambda_1, \ldots, \lambda_p)$ where λ_i, $i = 1, \ldots, p$ are the eigenvalues of $X'_* X_*$. The model in Eq. (1.13) may be written as

$$y_* = Z\gamma + u_* \tag{2.1}$$

where $Z = X_* G$, $\gamma = G'\beta$ and $Z'Z = G'X'_* X_* G = \Lambda$. The OLS estimator of γ is given by $\hat{\gamma} = (Z'Z)^{-1}Z'y_*$ which is unbiased and its variance-covariance matrix is $Var(\hat{\gamma}) = \sigma^2 (Z'Z)^{-1}$. Also, the corresponding ridge estimator is

$$\hat{\gamma}_k = (Z'Z + kI)^{-1} Z'y_* = [I - k(Z'Z + kI)^{-1}]\hat{\gamma}. \tag{2.2}$$

To circumvent the problem of bias in ridge estimator, Kadiyala[11] proposed a class of almost unbiased estimators which included the bias corrected ridge estimator but it was not operational due to the involvement of unknown parameters. Later Ohtani[20] replaced the unknown parameters by their ridge estimator and made it feasible. The use of the Jackknife procedure to reduce the bias of the ridge estimator and the properties of the Jackknifed ridge estimator were first studied by Singh et al.[25]. Jackknife technique was introduced by Quenouille[22] as a method for reducing the bias of estimator. Later Tukey[29] proposed that this technique may also offer a simple method to obtain the confidence intervals for the parameters of interest. A large number of papers appeared dealing with large-sample properties and empirical validations in common applications like estimation of variances, correlations and ratios. Jackknife procedure was initially applied to balanced situations i.e., equal sample sizes, equal variances, etc. (see Miller[15,16] and Hinkley[7]).

Following Singh et al.[25], we can directly write the Jackknifed form of $\hat{\gamma}_k$ as

$$\hat{\gamma}_J = [I - k^2 (Z'Z + kI)^{-2}]\hat{\gamma}. \tag{2.3}$$

Since $\beta = G\gamma$ and $Z'Z = G'X'_* X_* G = G'(X'V^{-1}X + R'W^{-1}R)G$, the Jackknifed estimator of β for model in Eq. (1.1) with restriction in Eq. (1.6) and assumption in Eq. (1.7) is given by

$$\hat{\beta}_J = G\hat{\gamma}_J = [I - k^2 (X'V^{-1}X + R'W^{-1}R + kI)^{-2}]\hat{\beta}_m. \tag{2.4}$$

We call this as Jackknifed SRR (JSRR) estimator. Next section consists of the comparison between SRR and Jackknifed SRR estimator on the basis of bias and mean square error.

Remark 2.1. Recently Khurana *et al.*[12] have shown that the same form of the Jackknifed ridge estimator of β can be obtained directly without transformation.

3. Comparison between SRR and Jackknifed SRR Estimators

Theorem 3.1. *Consider any $k > 0$. Then the difference*

$$D = \sum |Bias(\hat{\beta}_k)|_i^2 - |Bias(\hat{\beta}_J)|_i^2 > 0.$$

Proof. For component wise comparison, we make use of the canonical form defined in Eq. (2.1):

$$\hat{\gamma}_k = [I - k(Z'Z + kI)^{-1}]\hat{\gamma};$$

$$Bias(\hat{\gamma}_k) = E(\hat{\gamma}_k) - \gamma = -k(Z'Z + kI)^{-1}\gamma;$$

$$\hat{\gamma}_J = [I - k^2(Z'Z + kI)^{-2}]\hat{\gamma};$$

$$Bias(\hat{\gamma}_J) = E(\hat{\gamma}_J) - \gamma = -k^2(Z'Z + kI)^{-2}\gamma.$$

On comparing $|Bias(\hat{\gamma}_k)|_i$ with $|Bias(\hat{\gamma}_J)|_i$ where $|.|_i$ denotes the absolute value of the i-th component, we have

$$|Bias(\hat{\gamma}_k)|_i - |Bias(\hat{\gamma}_J)|_i = \frac{k}{\lambda_i + k}|\gamma_i| - \frac{k^2}{(\lambda_i + k)^2}|\gamma_i|$$

$$= \frac{\lambda_i k}{(\lambda_i + k)^2}|\gamma_i|$$

which is a positive quantity (see also Singh *et al.*[25] and Nomura[19]) .

Also, if we look at the expression of total bias in terms of original parameter vector β, we find that

$$D = \sum \{|Bias(\hat{\beta}_k)|_i^2 - |Bias(\hat{\beta}_J)|_i^2\}$$

$$= \beta' \{G[k^2(Z'Z + kI)^{-2} - k^4(Z'Z + kI)^{-4}]G'\} \beta \qquad (3.1)$$

where use has been made of the fact that $Bias(\hat{\beta}) = GBias(\hat{\gamma})$. Since the matrix in braces in R.H.S of Eq. (3.1) is positive definite, $D > 0$. This proves the theorem (see also Singh *et al.*[25]). $\square$

Lemma 3.1. [Farebrother[2]] Let A be a positive definite matrix, β be a $p \times 1$ vector. Then $A - \beta\beta'$ is a nonnegative definite matrix if and only if $\beta' A^{-1} \beta \leq 1$ is satisfied.

Theorem 3.2. *Consider $k > 0$. Then the difference*

$$\Delta = MSE(\hat{\beta}_J) - MSE(\hat{\beta}_k)$$

is a positive definite matrix if and only if the following inequality is satisfied

$$\beta'\{L^{-1}[\sigma^2 H + k^4 B^{-2}\beta\beta' B^{-2}]L^{-1}\}^{-1}\beta \leq 1$$

where $L = B^{-1}kI_p$ and $B = (X'V^{-1}X + R'W^{-1}R + kI_p)$.

Proof. We can express $\hat{\beta}_k$ as follows:

$$\hat{\beta}_k = (X'V^{-1}X + R'W^{-1}R + kI_p)^{-1}(X'V^{-1}y + R'W^{-1}r), \quad k > 0$$
$$= B^{-1}S\hat{\beta}_m = [I - B^{-1}kI_p]\hat{\beta}_m$$

where $B = X'V^{-1}X + R'W^{-1}R + kI_p$ and $S = X'V^{-1}X + R'W^{-1}R$.
Also note that

$$Bias(\hat{\beta}_k) = -kB^{-1}\beta \tag{3.2}$$

and

$$Var(\hat{\beta}_k) = [I - B^{-1}kI_p]Var(\hat{\beta}_m)[I - B^{-1}kI_p]'$$
$$= [I - B^{-1}kI_p]\sigma^2 S^{-1}[I - B^{-1}kI_p]' \tag{3.3}$$

where the fact that $Var(\hat{\beta}_m) = \sigma^2 S^{-1}$ is used. Thus from equations (3.3) and (3.2) we have

$$MSE(\hat{\beta}_k) = [I - B^{-1}kI_p]\sigma^2 S^{-1}[I - B^{-1}kI_p]' + k^2 B^{-1}\beta\beta' B^{-1}. \tag{3.4}$$

Similarly, we can write for $\hat{\beta}_J$

$$\hat{\beta}_J = [I - (X'V^{-1}X + R'W^{-1}R + kI_p)^{-2}k^2]\hat{\beta}_m = [I - k^2 B^{-2}]\hat{\beta}_m, \quad k > 0$$

$$Bias(\hat{\beta}_J) = -k^2 B^{-2}\beta \tag{3.5}$$

and

$$Var(\hat{\beta}_J) = [I - k^2 B^{-2}]\sigma^2 S^{-1}[I - k^2 B^{-2}]'. \tag{3.6}$$

Thus from Eq. (3.6) and Eq. (3.5), we get

$$MSE(\hat{\beta}_J) = [I - k^2 B^{-2}]\sigma^2 S^{-1}[I - k^2 B^{-2}]' + k^4 B^{-2}\beta\beta' B^{-2}. \tag{3.7}$$

Using Eq. (3.4) and Eq. (3.7), Δ becomes

$$\Delta = \sigma^2 H + k^4 B^{-2} \beta \beta' B^{-2} - k^2 B^{-1} \beta \beta' B^{-1} \qquad (3.8)$$

where $H = [I - k^2 B^{-2}] S^{-1} [I - k^2 B^{-2}]' - [I - kB^{-1}] S^{-1} [I - kB^{-1}]'$ which can be easily shown to be a positive definite matrix (for detailed proof, see Theorem 3 or Theorem 4 of Khurana et al.[12]). Now the difference Δ is positive definite if and only if $L^{-1} \Delta L^{-1}$ is positive definite.

$$L^{-1} \Delta L^{-1} = L^{-1} [\sigma^2 H + k^4 B^{-2} \beta \beta' B^{-2}] L^{-1} - \beta \beta'. \qquad (3.9)$$

From Eq. (3.9), we see that the matrix $[\sigma^2 H + k^4 B^{-2} \beta \beta' B^{-2}]$ is symmetric positive definite matrix. Therefore using Lemma 3.1, we conclude that $L^{-1} \Delta L^{-1}$ is positive definite if and only if following inequality holds

$$\beta' \{ L^{-1} [\sigma^2 H + k^4 B^{-2} \beta \beta' B^{-2}] L^{-1} \}^{-1} \beta \leq 1. \qquad \square$$

Remark 3.1. We see that $\Delta = 0$ if and only if $L^{-1} \Delta L^{-1} = 0$, which gives

$$L^{-1} [\sigma^2 H + k^4 B^{-2} \beta \beta' B^{-2}] L^{-1} = \beta \beta'$$

which is not true because the rank of left hand matrix is p and the rank of right hand matrix is either 0 or 1. So Δ can not be zero whenever $p > 1$.

Remark 3.2. Noting that the matrix $k^4 B^{-2} \beta \beta' B^{-2}$ in Eq. (3.8) is positive definite, a simple sufficient condition for Δ to be positive definite using Lemma 3.1 is given by

$$\sigma^{-2} \beta' L H^{-1} L \beta \leq 1.$$

From the above two theorems, we observe that JSRR estimator certainly improves upon SRR estimator in terms of bias but may not always perform better when it comes to mean square error. However, following Singh and Chaubey[26], we can get the condition where JSRR dominates SRR in terms of mean square error for transformed model. In the next section, we present a numerical example to validate our findings.

4. Numerical Illustration

In this section, we illustrate the theoretical results with a numerical example. We use the data on the weekly quantities of shampoos sold to determine the weekly variation of sales of the shampoos and to estimate the demand for future weekly sales. This example was originally used by Bayhan and Bayhan[1] (see also Özkale[21]). There are two independent variables namely weekly list prices of the shampoos (X_1) and weekly list prices of some brand of soap (X_2), dependent variable (y) denote weekly quantities of shampoos sold (For the data tables, please refer to Bayhan and Bayhan[1]).

Here, the matrix V is assumed to be known (see also Trenkler[28], Özkale[21]) but in practice it is rarely known, so an estimator $\hat{V}$ is used. In the literature, some of the estimators of V are given by Firinguetti[4], Trenkler[28], Bayhan and Bayhan[1], Judge et $al.$[10]. For our study, we have used the form of the matrix V given by Firinguetti[4] as follows:

$$V = \begin{pmatrix} 1 & \rho & \rho^2 & \ldots & \rho^{n-1} \\ \rho & 1 & \rho & \ldots & \rho^{n-2} \\ \cdot & \cdot & \cdot & & \cdot \\ \cdot & \cdot & \cdot & & \cdot \\ \cdot & \cdot & \cdot & & \cdot \\ \rho^{n-1} & \rho^{n-2} & \rho^{n-3} & \ldots & 1 \end{pmatrix}$$

where the terms in V are generated from an AR(1) process as follows:

$$\epsilon_i = \rho\epsilon_{i-1} + u_i, \ |\rho| < 1, \ i = 1, 2, \ldots, n \tag{4.1}$$

where $u_i \sim N(0, \sigma_u^2)$, $E(u_i, u_j) = 0$ for all $i \neq j$, $\sigma^2 = \sigma_u^2/1 - \rho^2$. Also, the estimate of ρ is given by

$$\hat{\rho} = \frac{\displaystyle\sum_{i=1}^{n-1} \hat{\epsilon_i}\hat{\epsilon_{i+1}}}{\displaystyle\sum_{i=1}^{n} \hat{\epsilon_i^2}} \tag{4.2}$$

where $\hat{\epsilon_i}$ are the elements of residuals $\hat{\epsilon} = y - X\hat{\beta}$.

Variables have been standardized before the calculations. Following Bayhan and Bayhan[1] and Özkale[21], testing for the autocorrelation in fresh data using Durbin-Watson (DW) test leads to the value of test statistic 0.3533 which indicates the presence of autocorrelation at 5% level of significance with the critical values $d_L = 0.95$ and $d_U = 1.54$ for $n = 15$. Further using the test given by Godfrey[5], Özkale[21] do not reject the null hypothesis which states that ϵ_i's are generated by an AR(1) process.

Bayhan and Bayhan[1] used the historical data to estimate the value of V to proclaim that it is known. To apply the stochastic linear restrictions in Eq. (1.6), a subset of observations (observations 48 and 49) is chosen from the standardized historical data as follows

$$R = \begin{bmatrix} 0.1450 & 0.1049 \\ 0.0077 & 0.1850 \end{bmatrix}; \; r = \begin{bmatrix} 0.1303 \\ 0.1380 \end{bmatrix}.$$

Also, matrix W is determined using the first two rows and columns of V (see Bayhan and Bayhan[1]):

$$W = \begin{bmatrix} 1 & 0.35333 \\ 0.35333 & 1 \end{bmatrix}.$$

The values of estimates of GLS, MR, SRR, and JSRR are given in Table 1. The values of scalar mean square error (SMSE; trace of mean square error matrix) in Table 1 are calculated using the estimator of σ^2 as $\hat{\sigma}^2 = (y - X\hat{\beta}_g)'V^{-1}(y - X\hat{\beta}_g)/(n - p) = 0.1701399$. Also the estimator of shrinkage parameter k is obtained by $\hat{k} = p\hat{\sigma}^2/\hat{\beta}_g'\hat{\beta}_g = 0.4177769$ and β is replaced by $\hat{\beta}_g$ (see Firinguetti[4]).

Table 1.   Estimated values of coefficients, Abias and SMSE

Estimators	β_1	β_2	Abias	SMSE
$\hat{\beta}_g$	0.90067917	-0.05725619	-	4.761327
$\hat{\beta}_m$	0.6379431	0.2068868	-	2.965632
$\hat{\beta}_k$	0.4384289	0.3857928	0.8414882	0.407399
$\hat{\beta}_J$	0.4714962	0.3730181	0.7395632	0.5175715

The eigen values of $X'V^{-1}X$ are 16.49135846 and 0.03581131 which results in the condition number (square root of ratio of maximal to minimal eigen values) 21.45943. This shows that fresh data set have strong multicollinearity. This condition number is smaller when calculated for $X'V^{-1}X + R'W^{-1}R$ and comes out be 16.94420. This shows that inducing the prior information on the coefficient vector reduces the degree of collinearity to some extent. However, in case of SRR and JSRR estimators, for getting the condition number, we have to find the eigen values of the same matrix which is $(X'V^{-1}X + R'W^{-1}R + kI_p)$. Taking $k = 0.4177769$, condition number for $(X'V^{-1}X + R'W^{-1}R + kI_p)$ comes out to be 5.970841 which is smaller than that of $X'V^{-1}X + R'W^{-1}R$.

From Table 1, we find that MR estimator has less SMSE value than that of GLS estimator and it also corrects the wrong sign problem in coefficients of GLS which may be a reason of multicollinearity. Also, JSRR estimator has reduced the bias of SRR estimator but is not improving in the case of SMSE. In Table 1, we have considered only a specific value of k which is

0.4177769. Therefore to see the effects of different values of k on bias and means square error, we make use of Fig. 1 and Fig. 2. Fig. 1 shows that bias of JSRR is less than that of SRR estimator for all the values of k in this range. From Fig. 2, it is clear that mean square error of SRR is lesser than that of JSRR until a point which may be somewhere near $k = 1.4$ after which we do not observe much difference in the mean square error of both the estimators.

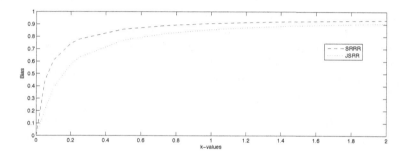

Fig. 1.   Bias comparison.

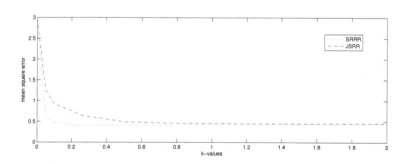

Fig. 2.   Mean square error comparison.

5. A Simulation Study

In the present section, our aim is to compare the performance of MRE, SRR and JSRR estimators on the basis of associated bias and mean square error with the help of Monte Carlo experiments.

The true model is $y = X\beta + u$ where $u \sim N(0, 1)$. Here β is taken as the normalized eigen vector corresponding to the largest eigen value of $X'V^{-1}X$ (see Newhouse and Oman[18], McDonald and Galarneau[14], Kibria[13]). The explanatory variables are generated from the following equation

$$x_{ij} = (1 - \gamma^2)^{\frac{1}{2}} w_{ij} + \gamma w_{ip}, \ i = 1, 2, \dots, n; \ j = 1, 2, \dots, p$$

where w_{ij} are independent standard normal pseudo-random numbers and γ^2 is the correlation between x_{ij} and $x_{ij'}$ for $j, j' < p$ and $j \neq j'$. When j or $j' = p$, the correlation will be γ. We have taken $\gamma = 0.7$ and 0.99 to investigate the effects of different degrees of collinearity with sample sizes $n = 15$ and 50 and $p = 3$. Four different values of ρ as 0.1, 0.4, 0.7, 0.9 have been used. For the restrictions, we have taken $r = X_0' y_0$, $R = X_0' X_0$ and $W = X_0' V X_0$ where X_0 is an $n \times p$ matrix which has columns as independent standard normal pseudo-random numbers. The feasible value of k is obtained by the optimal formula $k = \frac{p\sigma^2}{\beta'\beta}$ as given by Hoerl et al.[9], so that

$$\hat{k} = \frac{p\hat{\sigma}^2}{\hat{\beta}_m' \hat{\beta}_m}; \ \hat{\sigma}^2 = \frac{(y - X\hat{\beta}_m)'(y - X\hat{\beta}_m)}{n - p}.$$

This simulation study is patterned on that of McDonald and Galarnaeu[14]. For these different choices of ρ, γ and n, the experiment is replicated 5000 times. The average absolute bias and average mean square error are computed as follows:

$$Bias(\hat{\beta}_i) = \frac{1}{5000} \sum |\hat{\beta}_{ir} - \beta_i|,$$

and

$$MSE(\hat{\beta}) = \frac{1}{5000} \sum (\hat{\beta}_r - \beta)'(\hat{\beta}_r - \beta).$$

Here, $\hat{\beta}_{ir}$ denotes the estimate of $i - th$ parameter in $r - th$ replication and β_1, β_2 and β_3 are the true parameter values. Results of the simulation study are given in Tables 2 and 3. From Table 2, we see that the bias of JSRR is smaller than that of SRR in almost all the cases for different values of ρ and γ. From Table 3, we find that as the degree of collinearity gets larger, performance JSRR gets much better than that of MRE in the sense of mean square error. However, SRR is always better than the other two estimators in all the cases considered. It is also important to notice that when the degree of collinearity is small, there is very less difference in the mean square error of SRR and JSRR in comparison to others.

Table 2. Bias comparison.

		ρ=0.1			ρ=0.4		
n	γ	MRE	SRR	$JSRR$	MRE	SRR	$JSRR$
15	0.7	0.005189	0.031043	0.008023	0.004672	0.029940	0.006951
		0.001444	0.023294	0.004848	0.001260	0.027331	0.002972
		0.001642	0.034226	0.006406	0.001690	0.030697	0.005202
	0.99	0.006074	0.037259	0.010655	0.005926	0.037272	0.009987
		0.000083	0.027245	0.015405	0.000031	0.028070	0.009471
		0.000017	0.026039	0.010479	0.000329	0.026088	0.006986
50	0.7	0.001457	0.007178	0.002170	0.001049	0.002922	0.000498
		0.002342	0.007734	0.002814	0.002228	0.002055	0.001816
		0.000595	0.010038	0.000039	0.000570	0.009316	0.001128
	0.99	0.000892	0.006370	0.000210	0.000914	0.004922	0.001055
		0.000850	0.007369	0.001039	0.002831	0.001798	0.001602
		0.000140	0.009359	0.000907	0.001598	0.008472	0.003466

		ρ=0.7			ρ=0.9		
n	γ	MRE	$SRRR$	$JSRR$	MRE	$SRRR$	$JSRR$
15	0.7	0.003574	0.011262	0.002480	0.002222	0.006349	0.002526
		0.001242	0.021766	0.003065	0.000966	0.007579	0.000475
		0.001206	0.017040	0.000341	0.000688	0.007358	0.001058
	0.99	0.005154	0.023869	0.007649	0.004333	0.009346	0.005959
		0.000836	0.019272	0.004155	0.001987	0.007535	0.001213
		0.000142	0.017435	0.003204	0.000031	0.006137	0.000933
50	0.7	0.000583	0.002375	0.000842	0.000246	0.000859	0.000300
		0.001739	0.004097	0.001951	0.000952	0.001883	0.000997
		0.000442	0.004600	0.000159	0.000188	0.001494	0.000132
	0.99	0.000661	0.003042	0.002886	0.000320	0.000075	0.002197
		0.003191	0.005309	0.004413	0.002812	0.001746	0.002017
		0.002237	0.002532	0.000001	0.002124	0.003830	0.003971

Table 3. MSE comparison.

		ρ=0.1			ρ=0.4		
n	γ	MRE	$SRRR$	$JSRR$	MRE	$SRRR$	$JSRR$
15	0.7	0.119921	0.106414	0.119532	0.094216	0.087522	0.094125
	0.99	0.145557	0.114278	0.144628	0.113276	0.096755	0.112899
50	0.7	0.035340	0.033270	0.035313	0.030854	0.029079	0.030828
	0.99	0.048492	0.043518	0.048394	0.050541	0.044721	0.050419

		ρ=0.7			ρ=0.9		
n	γ	MRE	$SRRR$	$JSRR$	MRE	$SRRR$	$JSRR$
15	0.7	0.052024	0.049685	0.051963	0.019314	0.018884	0.019308
	0.99	0.070390	0.062507	0.070119	0.038958	0.034340	0.038791
50	0.7	0.019242	0.018439	0.019236	0.007261	0.007130	0.007261
	0.99	0.050641	0.043896	0.050494	0.043076	0.036962	0.042920

58

6. Concluding Remarks

In this study, for reducing the bias in the SRR estimator, we proposed an estimator which is the jackknifed form of SRR. We proved that JSRR estimator improves upon SRR in terms of bias but may not always have smaller mean square error value than that of SRR, however the conditions on the biasing parameter k for JSRR to have smaller mean square error than SRR are similar to those in Singh and Chaubey[26] and can be easily obtained. The results have been supported using a real data example as well as using Monte Carlo simulation.

Acknowledgments

Y.P. Chaubey would like to acknowledge Natural Sciences and Engineering Research Council of Canada for the partial support of this research through a Discovery Research Grant. Mansi Khurana would like to thank Banasthali University in granting her leave to facilitate the access of Common Wealth visiting scholarship at Concordia University.

References

1. G. M. Bayhan and M. Bayhan, *Computers and Industrial Engineering* **34**, 413 (1998).
2. R. W. Farebrother, *Journal of Royal Statistical Society* **B38**, 248 (1976).
3. D. E. Farrar and R. R. Glauber, *The Review of Economics and Statistica* **49(1)**, 92 (1967).
4. L. Firinguetti, *Communications in Statistics–Simulation and Computation* **18(2)**, 673 (1989).
5. L. G. Godfrey, *Econometrica* **46(6)**, 1293 (1978).
6. J. Groβ, *Statistics and Probability Letters* **65**, 57 (2003).
7. D. V. Hinkley, *Technometrics* **19(3)**, 285 (1977).
8. A. E. Hoerl and R. W. Kennard, *Technometrics* **20**, 69 (1970).
9. A. E. Hoerl, R. W. Kennard and K. Baldwin, *Communications in Statistics–Theory and Methods* **4**, 105 (1975).
10. G. G. Judge, W. E. Griffiths, R. C. Hill, H. Lütkepohl, T. C. Lee, *The Theory and Practice of Econometrics* (John Wiley and Sons, New York, 1985).
11. K. Kadiyala, *Economics Letters* **16**, 293 (1984).
12. M. Khurana, Y.P. Chaubey and S. Chandra, *Technical Report, #01/12*, Department of Mathematics and Statistics, Concordia University (2012). Available at http://spectrum.library.concordia.ca/974163/
13. B. M. G. Kibria, *Communications in Statistics–Simulation and Computation* **32(2)**, 419 (2003).
14. G. C. McDonald and D. I. Galarneau, *Journal of the American Statistical Association* **70**, 407 (1975).

15. R. G. Miller, *Biometrika* **61**, 1 (1974).
16. R. G. Miller, *Annals of Statistics* **2**, 880 (1974).
17. A. L. Nagar and N. C. Kakwani, *Econometrica* **32**, 174 (1964).
18. J. P. Newhouse, S. D. Oman, *Rand Report*, No. R-716-PR, 1 (1971).
19. M. Nomura, *Communications in Statistics–Simulation and Computation* **17**, 729 (1988).
20. K. Ohtani, *Communications in Statistics–Theory and Methods* **15**, 1571 (1986).
21. M. R. Özkale, *Journal of Multivariate Analysis* **100**, 1706 (2009).
22. M. H. Quenouille, *Biometrika* **43**, 353 (1956).
23. C. R. Rao and H. Toutenburg, *Linear Models Least Squares and Alternatives* (Springer-Verlag, New York, 1995).
24. N. Sarkar, *Communications in Statistics–Theory and Methods* **21**, 1987 (1992).
25. B. Singh, Y. P. Chaubey and T. D. Dwivedi, *Sankhya* **B48**, 342 (1986).
26. B. Singh and Y. P. Chaubey, *Statistical Papers* **28**, 53 (1987).
27. H. Theil and A. S. Goldberger, *International Economic Review* **2**, 65 (1961).
28. G. Trenkler, *Journal of Econometrics* **25**, 179 (1984).
29. J. W. Tukey, *Annals of Mathematical Statistics* **29**, 614 (1958).

MOLECULAR CLASSIFICATION OF ACUTE LEUKEMIA

G. CHEN and J. WU*

Department of Mathematics and Statistics, University of Calgary,
Calgary, AB T2N 1N4, Canada
* E-mail: jinwu@math.ucalgary.ca

The case-control studies are widely used in epidemiology to identify important indicators or factors. In this article, a two-sample semiparamtric model, which is equivalent to the logistic regression model for case-control data, is proposed to describe the gene expression level for acute lymphoblastic leukemia (ALL) patients and acute myeloid leukemia (AML) patients. Considering a leukemia data set containing 38 bone marrow samples (27 ALL, 11 AML), we propose the minimum Hellinger distance estimation (MHDE) for the underlying semi-parametric model and compare the results with those based on the classical maximum likelihood estimation (MLE). Based on the MHDE and MLE, Wald tests of significance are carried out to select marker genes. Further, using the idea of minimizing the sum of weighted misclassification rates, we develop a new classification rule based on the selected marker genes. To test our proposed classification rule, another independent leukemia data set (20 ALL, 14 AML) is analyzed and the result shows that 31 out of 34 patients are successfully classified. In the training data set, 36 out of 38 are successfully classified.

1. Introduction

Leukemia is a type of cancer of the blood or bone marrow characterized by an abnormal increase of abnormal blood cells. There are two different ways to classify leukemia. According to the type of stem cell leukemia develops from, it can be divided into Myelogenous (or myeloid) leukemia or Lymphocytic (or lymphoblastic) leukemia. According to how quickly the leukemia develops and grows, it can be classified as either acute leukemia or chronic leukemia. Therefore, there are mainly four types of leukemia: *acute lymphoblastic leukemia* (ALL), *acute myelogenous leukemia* (AML), *chronic lymphocytic leukemia* (CLL) and *chronic myelogenous leukemia* (CML).

It is critical for a doctor to diagnose which type of leukemia each patient has so that a treatment that works best for that type can be exercised. For example, a high-risk patient might need bone marrow transplantation immediately while a low-risk patient usually starts with standard therapy.

Although treatments for one type of leukemia sometimes can be applied to another as well, the treatment effect is usually diminished sharply. Therefore, leukemia classification is critical for a successful treatment. To classify leukemia, an experienced hematopathologist's interpretation of the tumor's morphology, histochemistry, immunophenotyping, and cytogenetic analysis is needed in current clinical practice.

Over the last few decades, gene expression has been receiving increasing attention simply due to the fact that its location, timing and amount (level) have a profound effect on how human body behaves. Fortunately, the technique of DNA microarrays can monitor the expression levels of a large amount of genes simultaneously. Based on gene expression monitoring by DNA microarrays, we are trying to develop an accurate and systematic statistical approach to classify leukemia.

The remainder of this paper is organized as follows. In Section 2, we briefly introduce the leukemia data set to be analyzed. In Section 3, a two-sample semiparametric model is proposed to model the gene expression levels, for which two methods of estimation, namely, minimum Hellinger distance estimation (MHDE) and maximum likelihood estimation (MLE), are developed and discussed. In Section 4, based on Wald test statistics, a procedure is proposed to select marker genes. In Section 5, we construct a leukemia classification rule based on the marker genes and then apply the rule to another independent leukemia data set. Finally, some discussions and remarks are given in Section 6.

2. Data Description

The leukemia data was first published in Golub *et al.*[3] and contains two subsets: training data set and independent data set. There are 38 acute leukemia patients in the training data set, among which 27 are ALL and 11 are AML, and all of them are bone marrow (BM) samples. The independent data set contains 34 acute leukemia patients with 20 ALL and 14 AML. Unlike the training data set, the specimens consist of 24 bone marrow samples and 10 peripheral blood (PB) samples.

For each individual, 7129 observations were generated from Affymetrix chips under microarray technology framework. Experimental protocol, rescaling factors and other details can be found from the website http://www.genome.wi.edu.mit.edu/MPR. After excluding some observations designed to control the quality of scanned microarray image, each individual contains probes for 6129 human genes. A quantitative expression level is available for each gene. All the data were obtained at the time of leukemia diagnosis.

To see the pattern of the gene expression levels for both ALL group and AML group, we give in Figure 1 the boxplots of the expression levels for some significant[a] genes based on the training data set. Clearly, outliers are present and some of them are quite extreme. Therefore, resistance to outliers is a desired property for any reliable statistical inference. In the following sections, we will develop robust procedures to analyze the leukemia data. Since the above described leukemia data only contains the information for acute leukemia classification, we will focus in this paper on how to classify acute leukemia patients into either ALL or AML.

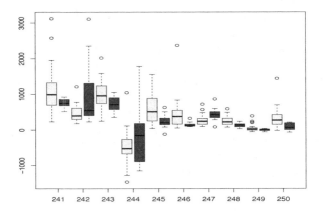

Fig. 1. Boxplot of gene expression level for 241th-250th significant genes with yellow for ALL group and green for AML group.

3. Methodology

In this section, we will propose a two-sample semiparametric model to describe the gene expression levels. For the proposed model, we will construct both the MLE and MHDE.

---

[a]Detailed procedure for selecting significant genes will be discussed in Section 4.

3.1. *A two-sample semiparametric model*

The leukemia classification (ALL or AML) is simply a problem of estimating a binary response, so intuitively one could use the logistic regression model. Let Y be a binary response variable and X be the associated $p \times 1$ covariate vector. Then the logistic regression model is given by

$$P(Y = 1|X = x) = \frac{\exp[\alpha^* + \beta^T r(x)]}{1 + \exp[\alpha^* + \beta^T r(x)]} \quad (1)$$

where $r(x) = (r_1(x), \ldots, r_p(x))^T$ is a $p \times 1$ vector of functions of x, α^* is a scalar parameter and β is a $p \times 1$ parameter vector. In most applications, $r(x) = x$ or $r(x) = (x, x^2)$. In the leukemia data, the covariate X is the gene expression level and we use $Y = 0$ (control) for classification result ALL and $Y = 1$ (case) for AML. Note that the leukemia data, like any other case-control data, are collected retrospectively in the sense that for individuals having $Y = 0$ (ALL) and having $Y = 1$ (AML), the values of the gene expression level X are observed. Specifically, let $X_1, \ldots, X_n$ be a random sample from a population with density function $g(x) = f(x|Y = 0)$ and, independent of the X_i's, let $Z_1, \ldots, Z_m$ be another random sample from a population with density function $h(x) = f(x|Y = 1)$. If $\pi = P(Y = 1)$, then (1) and Bayes's rule give the following two-sample semiparametric model:

$$\begin{aligned} X_1, \ldots, X_n &\stackrel{\text{i.i.d.}}{\sim} g(x) \\ Z_1, \ldots, Z_m &\stackrel{\text{i.i.d.}}{\sim} h(x) = g(x) \exp[\alpha + \beta^T r(x)], \end{aligned} \quad (2)$$

where α is a normalizing parameter that makes $g(x) \exp[\alpha + \beta^T r(x)]$ a density function. Models (1) and (2) are equivalent with parameters related by $\alpha = \alpha^* + \log[(1 - \pi)/\pi]$, which has been shown in Qin and Zhang.[5] The two unknown density functions $g(x)$ and $h(x)$ are linked by a weight function $\exp[\alpha + \beta^T r(x)]$. Now the problem has been transformed to estimating α and β treating g as a nuisance parameter.

For the leukemia data, g and h in model (2) are the distributions of gene expression level for ALL and AML, respectively. The X_i's are the gene expression levels in ALL and the Z_i's are those in AML. The benefit of model (2), compared with model (1), is that it describes the distributions of the gene expression levels for case and control groups separately and how they are linked. At the same time, one can easily transform α back to α^* and calculate $P(Y = 1|X = x)$ for classification purpose. Next, we will use both MLE and MHDE to estimate model (2).

3.2. *Maximum likelihood estimation*

Let G and H denote the distribution functions corresponding to g and h respectively. Denote $\{T_1, \ldots, T_N\}$ the pooled sample $\{X_1, \ldots, X_n; Z_1, \ldots, Z_m\}$ with $N = n + m$. The MLE is obtained by maximizing

$$L(\alpha, \beta, G) = \prod_{i=1}^{n} dG(X_i) \prod_{j=1}^{m} w(Z_j) dG(Z_j) = \left\{ \prod_{i=1}^{N} p_i \right\} \left\{ \prod_{j=1}^{m} w(Z_j) \right\}$$

subject to

$$p_i \geq 0, \quad \sum_{i=1}^{N} p_i = 1, \quad \sum_{i=1}^{N} p_i \{w(T_i) - 1\} = 0,$$

where $w(x) = \exp[\alpha + \beta^T r(x)]$ and $p_i = dG(T_i)$, $i = 1, \ldots, N$. The first constraint ensures p_i being non-negative and the second and third constraints make G and H both cumulative distribution functions. Zhang[7] has shown that the maximum value of L is attained at

$$\tilde{p}_i = \frac{1}{n} \frac{1}{1 + \rho \exp[\tilde{\alpha} + \tilde{\beta}^T r(T_i)]},$$

where $\rho = m/n$ and $(\tilde{\alpha}, \tilde{\beta})$ is the maximum semi-parametric likelihood estimator of (α, β) obtained as a solution to the following system of score equations:

$$\frac{\partial l(\alpha, \beta)}{\partial \alpha} = m - \sum_{i=1}^{N} \frac{\rho \exp[\alpha + \beta^T r(T_i)]}{1 + \rho \exp[\alpha + \beta^T r(T_i)]} = 0,$$

$$\frac{\partial l(\alpha, \beta)}{\partial \beta} = \sum_{j=1}^{m} r(Z_j) - \sum_{i=1}^{N} \frac{\rho \exp[\alpha + \beta^T r(T_i)]}{1 + \rho \exp[\alpha + \beta^T r(T_i)]} r(T_i) = 0,$$

with $l(\alpha, \beta)$ the profile log-likelihood function given by

$$l(\alpha, \beta) = \sum_{j=1}^{m} [\alpha + \beta^T r(T_j)] - \sum_{i=1}^{N} \log\{1 + \rho \exp[\alpha + \beta^T r(T_i)]\} - N \log n.$$

Zhang[7] has shown that the maximum semiparametric likelihood estimator $(\tilde{\alpha}, \tilde{\beta})$ is $\sqrt{n}$-consistent and asymptotically normally distributed.

3.3. *Minimum Hellinger distance estimation*

MLE has been proved to be most efficient, but some of the practical deficiencies of MLE are the lack of resistance to outliers and the general non-robustness with respect to (w.r.t.) model misspecification. As shown in Section 2, the leukemia data contains outliers. To overcome this problem, we introduce the method of MHDE that has been proved to have excellent robustness properties such as the resistance to outliers and robustness w.r.t. model misclassification (see Beran[1] and Donoho and Liu[2]).

Hellinger distance quantifies the similarity between two functions. The Hellinger distance between two density functions f_1 and f_2 is defined as

$$H(f_1, f_2) = \| f_1^{1/2} - f_2^{1/2} \|$$

where $\| \cdot \|$ denotes the L_2-norm. Suppose a simple random sample is from a population with density function h_θ with $\theta \in \Theta$ the only unknown parameter. Then the MHDE of θ under the parametric model $\{h_\theta : \theta \in \Theta\}$ is defined as

$$\bar{\theta} = \arg\min_{\theta \in \Theta} \| h_\theta^{1/2} - \hat{h}^{1/2} \|, \tag{3}$$

where $\hat{h}$ is an appropriate nonparametric estimator, e.g. kernel estimator, of the underlying true density h_θ based on the sample. Beran (1977) has shown that the MHDE defined in (3) is asymptotically efficient and exhibits good robustness properties.

For the two-sample semiparametric model (2), $h_\theta(x) = h(x) = g(x) \exp[\alpha + \beta^T r(x)]$ with $\theta = (\alpha, \beta^T)^T$. Unfortunately, besides θ, g in h_θ is also unknown. Intuitively, one can use a nonparametric estimator of g based on X_i's to replace g and then get an estimated h_θ. Specifically, we define kernel density estimators of g and h_θ as

$$g_n(x) = \frac{1}{nb_n} \sum_{i=1}^{n} K_0\left(\frac{x - X_i}{b_n}\right), \tag{4}$$

$$h_m(x) = \frac{1}{mb_m} \sum_{j=1}^{m} K_1\left(\frac{x - Z_j}{b_m}\right), \tag{5}$$

where K_0 and K_1 are density functions (kernels), bandwidths b_n and b_m are positive constants such that $b_n \to 0$ as $n \to \infty$ and $b_m \to 0$ as $m \to \infty$. Using the plug-in rule, for any $\theta \in \Theta$, we define $\widehat{h}_\theta$ as an estimator of h_θ:

$$\widehat{h}_\theta(x) = \exp[\alpha + \beta^T r(x)] g_n(x).$$

Now our proposed MHDE $\hat{\theta}$ of θ is the value which minimizes the Hellinger distance between the estimated model $\{\widehat{h}_\theta : \theta \in \Theta\}$ and the nonparametric density estimator h_m, i.e.

$$\hat{\theta} = \arg\min_{\theta \in \Theta} \| \widehat{h}_\theta^{1/2} - h_m^{1/2} \| . \tag{6}$$

Note that we do not impose constraint $\int \widehat{h}_\theta(x)dx = 1$ on θ. The reason being that, even for $\theta \in \Theta$ such that $\widehat{h}_\theta$ is not a density, it could make h_θ a density. The true parameter value θ may not make $\widehat{h}_\theta$ a density, but it is not reasonable to exclude θ as the estimate $\hat{\theta}$ of itself.

This MHDE has been investigated in Wu, Karunamuni and Zhang.[6] They have shown that the MHDE defined in (6) is $\sqrt{n}$-consistent and asymptotically normally distribute, and at the same time, is very robust to outliers and model misspecification.

For the nonparametric density estimators g_n and h_m defined in (4) and (5), respectively, we use Epanechnikov kernel function

$$K(u) = \begin{cases} \dfrac{3}{4}(1 - u^2), & \text{if } -1 < u < 1 \\ \\ 0, & \text{elsewhere} \end{cases}$$

for both K_0 and K_1, and use the data-driven bandwidths given by

$$b_n = 1.1926 \, \text{med}_i\big(\text{med}_j(|X_i - X_j|)\big) \cdot cn^{-2/5},$$

$$b_m = 1.1926 \, \text{med}_i\big(\text{med}_j(|Z_i - Z_j|)\big) \cdot cm^{-2/5}.$$

where c is an appropriate constant. Giving that the observations are relatively large in magnitude, we take $c = 10$ throughout our numerical studies. These choices of bandwidths guarantee the $\sqrt{n}$ convergence rate of the MHDE $\hat{\theta}$ (Wu, Karunamuni and Zhang[6]).

To ease the calculation, an algorithm, based on the Newton-Raphson method, is provided here to find the MHDE. For simplicity, we consider $r(x) = x$ and the following procedure could be easily adapted to other choices of $r(x)$. Note that the squared Hellinger distance can be reorganized as

$$D := \| \widehat{h}_\theta^{1/2} - h_m^{1/2} \|^2$$
$$= \int e^{(\alpha + \beta x)} g_n(x)dx - 2 \int e^{(\frac{1}{2}\alpha + \frac{1}{2}\beta x)} g_n^{1/2}(x) h_m^{1/2}(x)dx + 1.$$

Then, the first and second derivatives w.r.t α and β can be obtained as

$$\frac{\partial D}{\partial \alpha} = \int e^{(\alpha + \beta x)} g_n(x)dx - \int e^{(\frac{1}{2}\alpha + \frac{1}{2}\beta x)} g_n^{1/2}(x) h_m^{1/2}(x)dx$$

$$\frac{\partial D}{\partial \beta} = \int x e^{(\alpha+\beta x)} g_n(x) \mathrm{d}x - \int x e^{(\frac{1}{2}\alpha+\frac{1}{2}\beta x)} g_n^{1/2}(x) h_m^{1/2}(x) \mathrm{d}x$$

$$\frac{\partial^2 D}{\partial \alpha^2} = \int e^{(\alpha+\beta x)} g_n(x) \mathrm{d}x - \frac{1}{2} \int e^{(\frac{1}{2}\alpha+\frac{1}{2}\beta x)} g_n^{1/2}(x) h_m^{1/2}(x) \mathrm{d}x$$

$$\frac{\partial^2 D}{\partial \beta^2} = \int x^2 e^{(\alpha+\beta x)} g_n(x) \mathrm{d}x - \frac{1}{2} \int x^2 e^{(\frac{1}{2}\alpha+\frac{1}{2}\beta x)} g_n^{1/2}(x) h_m^{1/2}(x) \mathrm{d}x$$

$$\frac{\partial^2 D}{\partial \alpha \partial \beta} = \frac{\partial^2 D}{\partial \beta \partial \alpha} = \int x e^{(\alpha+\beta x)} g_n(x) \mathrm{d}x - \frac{1}{2} \int x e^{\frac{1}{2}(\alpha+\beta x)} g_n^{1/2}(x) h_m^{1/2}(x) \mathrm{d}x$$

Let (α_0, β_0) be the initial value, then by Newton-Raphson method, the MHDE can be updated by

$$\begin{pmatrix} \alpha_{k+1} \\ \beta_{k+1} \end{pmatrix} = \begin{pmatrix} \alpha_k \\ \beta_k \end{pmatrix} - \left[\begin{pmatrix} \frac{\partial^2 D}{\partial \alpha^2} & \frac{\partial^2 D}{\partial \alpha \partial \beta} \\ \frac{\partial^2 D}{\partial \beta \partial \alpha} & \frac{\partial^2 D}{\partial \beta^2} \end{pmatrix}^{-1} \begin{pmatrix} \frac{\partial D}{\partial \alpha} \\ \frac{\partial D}{\partial \beta} \end{pmatrix} \right] \Bigg|_{\alpha=\alpha_k, \beta=\beta_k} , \quad k = 0, 1, 2, \ldots$$

until it converges. Then the last iterated value is the MHDE. Throughout our numerical studies, we use the MLE as the initial value for MHDE. Given an appropriate initial value, one-step method is also feasible (Karunamuni and Wu[4]).

4. Marker Gene Selection

How could one use an initial collection of samples belonging to known classes (such as AML and ALL) to create a "class predictor" to classify new, unknown samples? In this section, a procedure will be constructed to select significant (marker) genes.

Since each individual has more than 6000 human genes, it would become a time-consuming and tedious task if all genes were included in our classification model. Meanwhile, it might bring plenty of noise to the model and hence deafen us to identify the type of leukemia correctly. Intuitively, we should only include those genes which have significantly different expression level between ALL and AML in our model. So we need to examine each human gene to see whether the expression pattern shows a strong difference between ALL and AML.

Let g_k and h_k be the distributions of the k-th gene expression level for ALL and AML respectively. Then g_k and h_k follow model (2) with $h_k = h_{\theta_k}$ and $\theta_k = (\alpha_k, \beta_k^T)^T$. If there is no difference in the k-th gene expression levels between ALL and AML, then $\theta_k = 0$. In this case, the k-th gene provides no information and thus should be excluded in the classification procedure. Note that α is only a normalizing parameter which won't change the shape of the distribution. Therefore, to decide whether the k-th gene is a marker gene or not, we only need to test

$$H_0 : \beta_k = 0 \quad \text{v.s.} \quad H_1 : \beta_k \neq 0. \tag{7}$$

Intuitively, we use the Wald test statistic

$$W_k = \beta_{Nk}^T [\widehat{Var}(\beta_{Nk})]^{-1} \beta_{Nk}, \tag{8}$$

where β_{Nk} is an estimator of β_k. In our case, β_{Nk} is either the MLE $\tilde{\beta}_k$ or the MHDE $\hat{\beta}_k$. Since both estimators are proved asymptotically normally distributed, under H_0, W_k is distributed as χ^2 with p degrees of freedom.

The asymptotic variances of $\tilde{\beta}_k$ and $\hat{\beta}_k$ are provided in Zhang[7] and Wu, Karunamuni and Zhang,[6] respectively, and therefore these can be readily estimated using the plug-in method. However, bootstrap resampling technique is preferred and used here to provide the estimated variance. The reason is simply that the sample sizes of the Leukemia data are small.

With a significance level of 5%, we test the hypotheses (7) for each gene based on MLE and MHDE. It turns out that 549 marker genes are selected by using MLE and 667 marker genes are selected by using MHDE.

5. Classification

Each marker gene provides information, more or less, on being ALL or AML. One needs to combine together all the information provided by all the marker genes and then form a classification, either ALL or AML. In this section, we will construct a classification rule by using the weighted misclassification rate and then test it with another independent leukemia data set.

5.1. Classification rule

Suppose totally s marker genes are selected. Now look at the k-th marker gene. We use g_k and h_k to denote the distributions of the k-th marker gene's expression level for ALL and AML respectively. Suppose for a particular leukemia patient, the expression level of this k-th marker gene is x_k. We call it a *misclassification into ALL* at x_k if a patient with x_k was classified as ALL when actually the patient is AML. The probability of this event is called the *misclassification rate into ALL* at x_k and is denoted as p_{k0}. Similarly we can define *misclassification into AML* and p_{k1}, the *misclassification rate into AML* at x_k.

Using Figure 2, we will demonstrate a way to classify this leukemia patient into either ALL or AML by using the k-th marker gene alone. Suppose the relative positions of g_k (ALL) and h_k (AML) are like the first graph in Figure 2, i.e. h_k is on the right side of g_k. If we classify a patient with x_k or higher value as AML and a value lower than x_k as ALL, then the misclassification rate into AML, p_{k1}, is the yellow area and that into ALL, p_{k0}, is the green area. If the relative positions of g_k (ALL) and h_k (AML) are like the second graph in Figure 2, i.e. h_k is on the left side of g_k, then we classify a patient with x_k or lower value as AML and a value higher than x_k as ALL and the misclassification rate into AML and ALL are still respectively the yellow area and green area. No matter which case, to classify this patient, one only needs to compare the yellow area with the green area, i.e. compare p_{k1} with p_{k0}. If $p_{k1} \leq p_{k0}$, then this patient is classified as AML, otherwise ALL. We call the yellow area the *misclassification region into AML* at x_k and the green area the *misclassification region into ALL* at x_k.

Note that p_{ki}, $i = 0, 1$, are not available simply because g_k and h_k are unknown. However, we can obtain their nonparametric estimators g_{nk} and h_{mk} constructed in the same way as in (4) and (5). By using the plug-in rule, the estimated misclassification rates into ALL and AML are respectively

$$\hat{p}_{k0} = \int_{A_{k0}} g_{nk}(x) \, dx,$$

$$\hat{p}_{k1} = \int_{A_{k1}} h_{mk}(x) \, dx,$$

where A_{k0} is the misclassification region into ALL group (MRL) and A_{k1} is the misclassification region into AML group (MRM). As shown in Figure 2, the relative positions of ALL and AML are important when computing misclassification rates. When h_{mk} (AML) is on the right side of g_{nk} (ALL), $\hat{p}_{k0}$ is the left tail probability of h_{mk} with MRL $A_{k0} = (-\infty, x_k)$ and $\hat{p}_{k1}$ is the right tail probability of g_{nk} with MRM $A_{k1} = (x_k, +\infty)$. This scenario is illustrated in the first graph. When h_{mk} is on the left side of g_{nk}, as shown in the second graph, $\hat{p}_{k0}$ is the right tail probability of h_{mk} with MRL $A_{k0} = (x_k, +\infty)$ and $\hat{p}_{k1}$ is the left tail probability of g_{nk} with MRM $A_{k1} = (-\infty, x_k)$.

Now we combine the misclassification rates for all the s marker genes together and form a unique decision on classification.

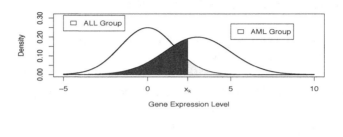

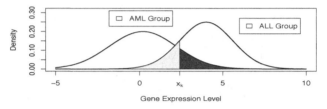

Fig. 2.   Classification Schematic Diagram.

Although all the marker genes passed significance test, they still have different significance levels. In other words, each maker gene contributes differently to the classification result. Therefore, different weights should be assigned to reflect their different contributions to the classification rule. Intuitively, we choose the relative size of the Wald test statistic as the weight:

$$w_k = \frac{|W_k|}{\sum_{i=1}^{s} |W_i|}, \quad k = 1, \ldots, s,$$

where W_k is the Wald test statistic defined in (8). Now we define the total misclassification rate (TMR) into ALL and AML, respectively, as

$$TMR_{ALL} = \sum_{k=1}^{s} w_k \hat{p}_{k0},$$

$$TMR_{AML} = \sum_{k=1}^{s} w_k \hat{p}_{k1}.$$

Clearly, the TMR is the weighted sum of all marker genes' misclassification rates. Now to give a unique classification decision based on all s marker genes, one can simply compare TMR_{ALL} with TMR_{AML}. A new acute leukemia patient should be classified to a group with lower misclassification rate. Define the ALL classification confidence (CC) coefficient as

$$CC = TMR_{AML} - TMR_{ALL} = \sum_{k=1}^{s} w_k(\hat{p}_{k1} - \hat{p}_{k0}).$$

It is easy to see that $-1 \le CC \le 1$. A value of CC closer to 1 signifies an ALL patient, however, that closer to -1 signifies an AML patient. Now our classification rule can be summarized as

Belongs to ALL group,   if $CC > 0$.
Undertermined,   if $CC = 0$.
Belongs to AML group,   if $CC < 0$.

5.2. Classification results and discussion

Classification results of independent set under MLE and MHDE are presented, respectively, in Tables 1 and 2, while for training set the results are presented in Tables 3 and 4.

For the independent set, the classification results are same no matter which estimation method is applied: 31 patients out of 34 are classified successfully. Also note that the same three patients are misclassified in both methods. All those three misclassified patients come from AML group and are misclassified as ALL. Among them, two patients belong to bone marrow sample and one of them corresponds to peripheral blood sample. Overall misclassification rate is 3/34=8.8%; misclassification rate for ALL patients is 0/20=0% while that for AML patients is 3/14=21.4%. Even though the classification results based on MHDE and MLE are exactly the same for each patient, with a closer examination of the CC values, we observe that the CC values of all three misclassified patients are closer to zero when MHDE is used than those when MLE is used, particularly for patient with ID 66. Note that for these three patients, negative CC values will give correct classification. Thus we expect that, in the long run, the MHDE will tend to give more accurate classification than MLE, even though we couldn't observe this for the small sample size here.

For training set, the classification results are same as well for MLE and MHDE method: 36 patients out of 38 are classified successfully and the same two patients are misclassified. Unlike for the independent set, the two misclassified patients come from ALL group and are misclassified as AML. Overall misclassification rate is 2/38=5.3%; misclassification rate for ALL patients is 2/27=7.4% while that for AML patients is 0/11=0%. As for the CC values, similar phenomenon is observed as for the independent set: the

CC values of all misclassified patients are closer to zero when MHDE is used than those when MLE is used. Hence it strengthens our belief that MHDE is preferred over MLE in the long run.

Table 1. Independent Set Classification Results based on MLE.

ID	CC	Predict	Actual	B/P	ID	CC	Predict	Actual	B/P
39	0.0301	ALL	ALL	BM	55	0.0578	ALL	ALL	BM
40	0.0393	ALL	ALL	BM	56	0.0943	ALL	ALL	BM
42	0.0236	ALL	ALL	BM	59	0.0816	ALL	ALL	BM
47	0.1355	ALL	ALL	BM	52	-0.1286	AML	AML	PB
48	0.4003	ALL	ALL	BM	53	-0.1071	AML	AML	BM
49	0.1036	ALL	ALL	BM	51	-0.2391	AML	AML	BM
41	0.2827	ALL	ALL	BM	50	-0.1995	AML	AML	BM
43	0.0818	ALL	ALL	BM	**54**	**0.1219**	**ALL**	**AML**	**BM**
44	0.1561	ALL	ALL	BM	57	-0.0981	AML	AML	BM
45	0.0941	ALL	ALL	BM	58	-0.1626	AML	AML	BM
46	0.1033	ALL	ALL	BM	60	-0.0012	AML	AML	BM
70	0.0331	ALL	ALL	PB	61	-0.0555	AML	AML	BM
71	0.0625	ALL	ALL	PB	65	-0.1153	AML	AML	BM
72	0.1353	ALL	ALL	PB	**66**	**0.0379**	**ALL**	**AML**	**BM**
68	0.3816	ALL	ALL	PB	63	-0.0096	AML	AML	PB
69	0.3274	ALL	ALL	PB	**64**	**0.0127**	**ALL**	**AML**	**PB**
67	0.0547	ALL	ALL	PB	62	-0.0358	AML	AML	PB

Table 2. Independent Set Classification Results based on MHDE.

ID	CC	Predict	Actual	B/P	ID	CC	Predict	Actual	B/P
39	0.0122	ALL	ALL	BM	55	0.0329	ALL	ALL	BM
40	0.0336	ALL	ALL	BM	56	0.0715	ALL	ALL	BM
42	0.0091	ALL	ALL	BM	59	0.0506	ALL	ALL	BM
47	0.1243	ALL	ALL	BM	52	-0.1841	AML	AML	PB
48	0.4002	ALL	ALL	BM	53	-0.0783	AML	AML	BM
49	0.0827	ALL	ALL	BM	51	-0.2215	AML	AML	BM
41	0.2632	ALL	ALL	BM	50	-0.1960	AML	AML	BM
43	0.0263	ALL	ALL	BM	**54**	**0.1115**	**ALL**	**AML**	**BM**
44	0.1417	ALL	ALL	BM	57	-0.1372	AML	AML	BM
45	0.0731	ALL	ALL	BM	58	-0.1732	AML	AML	BM
46	0.0922	ALL	ALL	BM	60	-0.0463	AML	AML	BM
70	0.0069	ALL	ALL	PB	61	-0.0886	AML	AML	BM
71	0.0507	ALL	ALL	PB	65	-0.1462	AML	AML	BM
72	0.1069	ALL	ALL	PB	**66**	**0.0032**	**ALL**	**AML**	**BM**
68	0.3650	ALL	ALL	PB	63	-0.0161	AML	AML	PB
69	0.3005	ALL	ALL	PB	**64**	**0.0108**	**ALL**	**AML**	**PB**
67	0.0085	ALL	ALL	PB	62	-0.0672	AML	AML	PB

Table 3.   Training Set Classification Results based on MLE.

ID	CC	Predict	Actual	B/P	ID	CC	Predict	Actual	B/P
1	0.3305	ALL	ALL	BM	20	0.6019	ALL	ALL	BM
2	0.2040	ALL	ALL	BM	21	0.3436	ALL	ALL	BM
3	0.3417	ALL	ALL	BM	**22**	**−0.0596**	**AML**	**ALL**	**BM**
4	0.2956	ALL	ALL	BM	23	0.1542	ALL	ALL	BM
5	0.5058	ALL	ALL	BM	24	0.4914	ALL	ALL	BM
6	0.1558	ALL	ALL	BM	25	0.0175	ALL	ALL	BM
7	0.1612	ALL	ALL	BM	26	0.2190	ALL	ALL	BM
8	0.0381	ALL	ALL	BM	27	0.0981	ALL	ALL	BM
9	0.3987	ALL	ALL	BM	34	-0.2667	AML	AML	BM
10	0.1596	ALL	ALL	BM	35	-0.2926	AML	AML	BM
11	0.2725	ALL	ALL	BM	36	-0.3349	AML	AML	BM
12	**−0.1565**	**AML**	**ALL**	**BM**	37	-0.4262	AML	AML	BM
13	0.5445	ALL	ALL	BM	38	-0.3438	AML	AML	BM
14	0.3579	ALL	ALL	BM	28	-0.1992	AML	AML	BM
15	0.5699	ALL	ALL	BM	29	-0.2522	AML	AML	BM
16	0.4089	ALL	ALL	BM	30	-0.3437	AML	AML	BM
17	0.4042	ALL	ALL	BM	31	-0.2667	AML	AML	BM
18	0.1694	ALL	ALL	BM	32	-0.1392	AML	AML	BM
19	0.2157	ALL	ALL	BM	33	-0.4108	AML	AML	BM

Table 4.   Training Set Classification Results based on MHDE.

ID	CC	Predict	Actual	B/P	ID	CC	Predict	Actual	B/P
1	0.3256	ALL	ALL	BM	20	0.5721	ALL	ALL	BM
2	0.1963	ALL	ALL	BM	21	0.3426	ALL	ALL	BM
3	0.3275	ALL	ALL	BM	**22**	**−0.0455**	**AML**	**ALL**	**BM**
4	0.2889	ALL	ALL	BM	23	0.1667	ALL	ALL	BM
5	0.4745	ALL	ALL	BM	24	0.4747	ALL	ALL	BM
6	0.1499	ALL	ALL	BM	25	0.0227	ALL	ALL	BM
7	0.1553	ALL	ALL	BM	26	0.2198	ALL	ALL	BM
8	0.0455	ALL	ALL	BM	27	0.1070	ALL	ALL	BM
9	0.3871	ALL	ALL	BM	34	-0.2547	AML	AML	BM
10	0.1672	ALL	ALL	BM	35	-0.2733	AML	AML	BM
11	0.2745	ALL	ALL	BM	36	-0.3310	AML	AML	BM
12	**−0.1391**	**AML**	**ALL**	**BM**	37	-0.4277	AML	AML	BM
13	0.5262	ALL	ALL	BM	38	-0.3277	AML	AML	BM
14	0.3318	ALL	ALL	BM	28	-0.1916	AML	AML	BM
15	0.5466	ALL	ALL	BM	29	-0.2446	AML	AML	BM
16	0.4039	ALL	ALL	BM	30	-0.3372	AML	AML	BM
17	0.3818	ALL	ALL	BM	31	-0.2606	AML	AML	BM
18	0.1576	ALL	ALL	BM	32	-0.1501	AML	AML	BM
19	0.2128	ALL	ALL	BM	33	-0.4127	AML	AML	BM

6. Further Discussion

Our study of the leukemia data shows that both MLE and MHDE are very promising in marker gene selection and classification. Further modifications and improvements will increase the accuracy and reliability. All numerical analysis in this paper is based on model (2) with $r(x) = x$. To make the model more reliable, one could add quadratic term. We are now continuing our study by adding the quadratic term, i.e. $r(x) = (x, x^2)$.

Besides the misclassification rates, we are also considering some other classification criteria. One intuitive way is to look at directly the chance of being ALL or AML based on each single marker gene. Since we have the estimation of parameters under model (2), we can easily get the estimation of parameters under model (1) according to the relationships between the two models. Then we can get the estimation for the chance of being ALL. With appropriate weights, we can use the weighted sum of chance of being ALL and that of being AML to create the classification rule.

Although a higher weight is assigned to a gene that is more significant, there is still too much noise in the model. One way to solve this problem is to lower the significance level and then reduce the number of marker genes. The second way is to test all the genes simultaneously. Third, double-weight could be used. For example, one could multiply the original weight by another penalty weight that is proportional to the reciprocal of minimized Hellinger distance (or maximized log-likelihood).

References

1. R. Beran, *Ann. Statist.* **5**, 445 (1977).
2. David L. Donoho and Richard C. Liu, *Ann. Statist.* **16**, 552 (1988).
3. T.R. Golub, D.K. Slonim, P. Tamayo, C. Huard, M. Gaasenbeek, J.P. Mesirov, H. Coller, M. Loh, J.R. Downing, M.A. Caligiuri, C.D. Bloomfield, and E.S. Lander, *Science* **286**, 531 (1999).
4. R.J. Karunamuni and J. Wu, *Computational Statistics and Data Analysis* **55**, 3148 (2011).
5. J. Qin and B. Zhang, *Biometrika* **84**, 609 (1997).
6. J. Wu, R.J. Karunamuni and B. Zhang, *Journal of Multivariate Analysis* **101**, 1102 (2010).
7. B. Zhang, *Bernoulli* **6**, 491 (2000).

A NOTE ON A NONPARAMETRIC TREND TEST AND ITS COMPETITORS

K. DEKA

Department of Statistics, Handique Girls' College,
Guwahati-781001, Assam, India
E-mail: kdeka2010@gmail.com

B. GOGOI

Department of Statistics, Dibrugarh University,
Dibrugarh-78600, Assam, India
E-mail: bipingogoi@gmail.com

One of the important problems in Statistics, often encountered in a medical and psychological applications is testing the equality of several means against ordered alternatives. The most popular nonparametric test in this context is the Jonckheere–Terpstra test. A new nonparametric test has been recently proposed by Bathke[5] that claims to be applicable for normal as well as non-normal populations. This paper provides an empirical comparison of the new test in contrast to the JT test and some parametric tests applicable to normal populations. Simulation results obtained here exhibit high power of the JT test in double exponential, logistic and Cauchy distributions, whereas test due to Bathke has the highest power for exponential populations.

1. Introduction

Testing equality of several means against ordered alternatives has been a standard problem in Statistics. Such problems are often encountered in medical and psychological experiments, where different groups are formed based on some covariate that may induce a trend amongst the group means. Bartholomew[2–4] developed likelihood ratio test for this situation and studied its asymptotic properties, when the observations are normally distributed. Terpstra[20] proposed a non-parametric test of equality of several means against an ordered alternative based on Kendall's τ that was later studied independently, in detail by Jonckheere.[17] This procedure is popularly known as the Jonckheere–Terpstra (JT) test (see Gibbons and Chakraborti[13]). Barlow *et al.*[1] proposed an alternative to Bartholomew's[2] test that assumes equality of variances of different populations. An adaptation of Welch's[21] method of adjusting the degrees of freedom has been studied by Büning and Kösler[12] for the Barlow *et al.*[1] test.

There have been many proposals in the direction of modifying these tests from various considerations during last four decades. Some research work has been done to find out the appropriate test out of various tests on the basis of the comparison of their asymptotic power function (e.g. Bartholomew,[3,4] Puri,[18] Berenson[6]) that are based on large sample theory. Some papers can be mentioned on comparison of power of the tests when the sample sizes are small (*e.g.* Jogdeo,[16] Ramsey,[19] Berenson and Gross,[7] Gogoi and Kakoty[14] and Büning[8–11]).

Recently, Bathke[5] has developed a test based on the idea of comparing the ranks of the sample observations obtained under the combined ranking with a trend variable. This test may allow for arbitrary underlying distributions including quantitative and discrete ordinal, or even binary data. This test was compared with the JT test and an asymptotically equivalent test given by Hettmansperger and Norton[15] based on weighted linear rank statistic, for normal and binomial populations.

Our purpose in this paper is to provide further study under other non-normal populations. Similar to the empirical study in Büning,[11] we consider the parametric test by Barlow *et al.*[1] and its adaptation to unequal variance case as proposed by Büning and Kössler[12] along with the JT test and Bathke's[5] test. We investigate through Monte Carlo simulation the actual levels of the tests in case of both equal and unequal variances and also their power for some symmetric and asymmetric distributions with short and long tails for detecting ordered alternatives in one-way layout, under variety of conditions. We offer suggestions for appropriate test selection under different situations.

The organization of the paper is as follows. Different tests are described in Sec. 2 and the simulation study is carried out in Sec. 3. Section 4 presents the results and the corresponding discussion where as Sec. 5 presents conclusions.

2. Model and the Tests

Here we consider a k-sample setup in the location-scale model frame work where the ith sample consists of n_i independent observations $X_{i1}, X_{i2}, ..., X_{in_i}; i = 1, 2, ..., k$. We let μ_i, σ_i denote the location and scale parameters, respectively, of the the ith population. Thus we assume that $X_{ij} \sim F_X((x - \mu_i)/\sigma_i), j = 1, 2, ..., n_i; i = 1, ..., k; -\infty < \mu_i < \infty, 0 <$

$\sigma_i < \infty$, where the distribution function $F_X(.)$ is assumed to be absolutely continuous. The problem now is to test $H_0 : \mu_1 = \mu_2 = ... = \mu_k$ vs. $H_0 : \mu_1 \le \mu_2 \le ... \le \mu_k$ with at least one strict inequality. Next we describe the tests we wish to consider for the above problem.

2.1. Parametric tests

For normal populations with equal variances, Barlow et al.[1] (see p. 184) test is based on the statistic T_{B_4} defined by

$$T_{B_4} = \frac{\sum_{i=1}^{k}(2i - k - 1)\bar{X}_i}{S\left(\sum_{i=1}^{k}\frac{(2i-k-1)^2}{n_i}\right)^{1/2}} \tag{1}$$

where

$$\bar{X}_i = \frac{1}{n_i}\sum_{j=1}^{n_i}X_{ij} \text{ and } S^2 = \frac{1}{N-k}\sum_{i=1}^{k}\sum_{j=1}^{n_i}(X_{ij} - \bar{X}_i)^2$$

with N denoting the total sample size $\sum_{i=1}^{k}n_i$.

Under normality of the populations, the null distribution of the test statistic B has Student's t-distribution with $N - k$ degrees of freedom. In the two sample case, it reduces to the usual t statistic for testing equality of the means of two normal populations with common variance. For the case of unequal variances, the modification of T_{B_4}, in the sense of Welch[21] (see Büning and Kösler,[12] is given by

$$T_{BW} = \frac{\sum_{i=1}^{k}(2i - k - 1)\bar{X}_i}{\left(\sum_{i=1}^{k}\frac{(2i-k-1)^2 S_i^2}{n_i}\right)^{1/2}} \tag{2}$$

where

$$S_i^2 = \frac{1}{n_i - 1}\sum_{j=1}^{n_i}(X_{ij} - \bar{X}_i)^2, \ i = 1, 2, ..., k.$$

The null distribution of the above test statistic may be approximated by a t- distribution with ν degrees of freedom that is given by

$$\nu = \frac{\left(\sum_{i=1}^{k}\frac{S_i^2}{n_i}\right)^2}{\sum_{i=1}^{k}\frac{S_i^4}{n_i^2(n_i-1)}}. \tag{3}$$

In practice, the value of ν is chosen to be the integer closest to it.

2.2. Nonparametric tests

Bathke's[5] test is based on the idea of comparing the ranks of a sample obtained under the combined ranking with a trend variable that has been proposed based on a unified approach that is applicable under a variety of situations. The test statistic in this case is given by

$$T_B = \frac{\sum_{i=1}^{k} \left(\sum_{l=1}^{i-1} n_l - \sum_{l=i+1}^{k} n_l \right) R_i}{\left(\sum_{i=1}^{k} \frac{n_i}{n_i-1} \left(\sum_{l=1}^{i-1} n_l - \sum_{l=i+1}^{k} n_l \right)^2 \sum_{j=1}^{n_i} (R_{ij} - \frac{N+1}{2})^2 \right)^{1/2}}, \quad (4)$$

where R_{ij} is the rank of X_{ij} in the combined sample and $R_i = \sum_{j=1}^{n_i} R_{ij}$. For equal sample sizes, $n_i \equiv n$, the above formula simplifies further to

$$T_B = \sqrt{\frac{n-1}{n}} \frac{\sum_{i=1}^{k}(i - \frac{k+1}{2})R_i}{\left(\sum_{i=1}^{k}(i - \frac{k+1}{2})^2 \sum_{j=1}^{n}(R_{ij} - \frac{N+1}{2})^2 \right)^{1/2}}. \quad (5)$$

We consider here the popular non-parametric test, known as the Jonckheere-Terpstra (henceforth denoted by JT) test that was described in the introduction for comparison purpose. The JT test is based on the sum over (see Gibbons and Chakraborti,[13] pp. 364-366) all Mann-Whitney statistics U_{il} for the two sample problem of comparing the samples i and l. The corresponding test-statistic is given by

$$T_{JT} = \sum_{i=1}^{k-1} \sum_{l=i+1}^{k} U_{il} = \sum_{i=1}^{k-1} \sum_{l=i+1}^{k} \sum_{r=1}^{n_i} \sum_{s=1}^{n_l} I(X_{ir} < X_{ls}). \quad (6)$$

The mean and variance are denoted by μ_{JT} and σ_{JT}^2, respectively, and are given by

$$\mu_{JT} = \frac{N - \sum_{i=1}^{k} n_i^2}{4},$$

and

$$\sigma_{JT}^2 = \frac{N^2(2N+3) - \sum_{i=1}^{k} n_i^2(2n_i+3)}{72}.$$

The null distribution of JT is asymptotically normal, and hence a large sample test based on this statistic may be based by comparing $(JT - \mu_{JT})/\sigma_{JT}$ with the right tail percentile of the standard normal distribution.

Table 1. Empirical size of the tests for different distributions for $k = 4$, sample-sizes: (10,10,10,10).

	Normal Samples											
α	0.10				0.05				0.01			
$\sigma-$ pattern	T_{B_4}	T_{BW}	T_{JT}	T_B	T_{B_4}	T_{BW}	T_{JT}	T_B	T_{B_4}	T_{BW}	T_{JT}	T_B
(i)	.0944	.0974	.0937	.0871	.0475	.0491	.0465	.0392	.0087	.0103	.0078	.0051
(ii)	.0979	.0989	.0966	.0905	.0466	.0493	.0454	.0406	.0087	.0115	.0087	.0051
(iii)	.1098	.1018	.1041	.1087	.0571	.0522	.0550	.0539	.0140	.0126	.0115	.0080
	Double-exponential Samples											
α	0.10				0.05				0.01			
$\sigma-$ pattern	T_{B_4}	T_{BW}	T_{JT}	T_B	T_{B_4}	T_{BW}	T_{JT}	T_B	T_{B_4}	T_{BW}	T_{JT}	T_B
(i)	.1012	.1045	.0960	.0923	.0510	.0597	.0461	.0433	.0110	.0105	.0080	.0051
(ii)	.1029	.1034	.0969	.0911	.0529	.0548	.0448	.0442	.0100	.0108	.0079	.0051
(iii)	.1148	.1092	.0997	.1076	.0600	.0556	.0500	.0522	.0126	.0120	.0090	.0160
	Logistic Samples											
α	0.10				0.05				0.01			
$\sigma-$ pattern	T_{B_4}	T_{BW}	T_{JT}	T_B	T_{B_4}	T_{BW}	T_{JT}	T_B	T_{B_4}	T_{BW}	T_{JT}	T_B
(i)	.1011	.1034	.0960	.0923	.0500	.0537	.0461	.0433	.0110	.0115	.0080	.0063
(ii)	.1029	.1045	.0976	.0925	.0529	.0553	.0458	.0432	.0100	.0126	.0082	.0065
(iii)	.1148	.1085	.1034	.1102	.0600	.0580	.0510	.0562	.0126	.0144	.0092	.0102
	Exponential Samples											
α	0.10				0.05				0.01			
$\sigma-$ pattern	T_{B_4}	T_{BW}	T_{JT}	T_B	T_{B_4}	T_{BW}	T_{JT}	T_B	T_{B_4}	T_{BW}	T_{JT}	T_B
(i)	.1009	.1025	.1006	.0858	.0511	.0561	.0535	.0413	.0113	.0136	.0095	.0059
(ii)	.1031	.1048	.1065	.1789	.0558	.0571	.0556	.1001	.0124	.0149	.0116	.0020
(iii)	.1171	.1070	.1159	.0806	.0643	.0579	.0658	.0691	.0170	.0180	.0163	.0303
	Cauchy Samples											
α	0.10				0.05				0.01			
$\sigma-$ pattern	T_{B_4}	T_{BW}	T_{JT}	T_B	T_{B_4}	T_{BW}	T_{JT}	T_B	T_{B_4}	T_{BW}	T_{JT}	T_B
(i)	.1187	.0809	.1000	.0880	.0393	.0263	.0470	.0390	.0039	.0014	.0089	.0054
(ii)	.1179	.0793	.1001	.0859	.0378	.0257	.0458	.0400	.0035	.0010	.0086	.0058
(iii)	.1176	.0840	.1000	.0948	.0379	.0247	.0516	.0452	.0039	.0010	.0086	.0072

Note: $\sigma-$ pattern: (i) $\sigma_i = 1, i = 1, 2, 3, 4$; (ii) $\sigma_1 = 1, \sigma_2 = 1.2, \sigma_3 = 1.3, \sigma_4 = 1.5$; (iii) $\sigma_1 = 1, \sigma_2 = 2, \sigma_3 = 3, \sigma_4 = 4$.

Table 2. Empirical size of the tests for different distributions for $k = 4$, sample-sizes: (10,15,20,25).

	Normal Samples											
α	0.10				0.05				0.01			
$\sigma-$ pattern	T_{B_4}	T_{BW}	T_{JT}	T_B	T_{B_4}	T_{BW}	T_{JT}	T_B	T_{B_4}	T_{BW}	T_{JT}	T_B
(i)	.1050	.1081	.0982	.0899	.0517	.0536	.0483	.0443	.0128	.0116	.0095	.0094
(ii)	.0801	.1014	.0933	.0940	.0234	.0510	.0480	.0475	.0059	.0107	.0095	.0106
(iii)	.0516	.1067	.0932	.1116	.0161	.0495	.0432	.0570	.0017	.0100	.0082	.0129

	Double-exponential Samples											
α	0.10				0.05				0.01			
$\sigma-$ pattern	T_{B_4}	T_{BW}	T_{JT}	T_B	T_{B_4}	T_{BW}	T_{JT}	T_B	T_{B_4}	T_{BW}	T_{JT}	T_B
(i)	.1082	.1036	.1018	.0913	.0486	.0510	.0525	.0431	.0109	.0102	.0090	.0073
(ii)	.0804	.1039	.1030	.0932	.0311	.0493	.0513	.0439	.0055	.0089	.0083	.0080
(iii)	.0485	.1025	.0982	.1021	.0152	.0486	.0468	.0545	.0015	.0096	.0085	.0186

	Logistic Samples											
α	0.10				0.05				0.01			
$\sigma-$ pattern	T_{B_4}	T_{BW}	T_{JT}	T_B	T_{B_4}	T_{BW}	T_{JT}	T_B	T_{B_4}	T_{BW}	T_{JT}	T_B
(i)	.1066	.1046	.1018	.0913	.0489	.0529	.0525	.0431	.0109	.0109	.0090	.0073
(ii)	.0839	.1031	.1014	.0932	.0329	.0500	.0507	.0452	.0050	.0091	.0083	.0081
(iii)	.0490	.1041	.0989	.1067	.0153	.0498	.0470	.0575	.0018	.0108	.0091	.0017

	Exponential Samples											
α	0.10				0.05				0.01			
$\sigma-$ pattern	T_{B_4}	T_{BW}	T_{JT}	T_B	T_{B_4}	T_{BW}	T_{JT}	T_B	T_{B_4}	T_{BW}	T_{JT}	T_B
(i)	.1081	.1010	.0995	.0813	.0514	.0532	.0509	.0401	.0122	.0147	.0095	.0070
(ii)	.0871	.1013	.0969	.2314	.0333	.0515	.0482	.1406	.0057	.0118	.0093	.0369
(iii)	.0525	.1010	.1018	.0895	.0177	.0532	.0513	.0820	.0021	.0147	.0092	.0564

	Cauchy Samples											
α	0.10				0.05				0.01			
$\sigma-$ pattern	T_{B_4}	T_{BW}	T_{JT}	T_B	T_{B_4}	T_{BW}	T_{JT}	T_B	T_{B_4}	T_{BW}	T_{JT}	T_B
(i)	.0993	.0879	.1015	.0880	.0595	.0276	.0517	.0390	.0145	.0023	.0124	.0052
(ii)	.0836	.0868	.1005	.0859	.0476	.0269	.0505	.0400	.0110	.0026	.0112	.0058
(iii)	.0519	.0862	.0937	.0948	.0237	.0269	.0468	.0452	.0034	.0027	.0083	.0072

Note: $\sigma-$ pattern: (i) $\sigma_i = 1, i = 1, 2, 3, 4$; (ii) $\sigma_1 = 1, \sigma_2 = 1.2, \sigma_3 = 1.3, \sigma_4 = 1.5$; (iii) $\sigma_1 = 1, \sigma_2 = 2, \sigma_3 = 3, \sigma_4 = 4$.

Table 3. Empirical power for 5% size tests for different distributions for $k = 4$, sample-sizes: (10,10,10,10).

$\sigma-$ pattern	(i)				(ii)				(iii)			
					Normal Samples							
$\mu-$ pattern	T_{B_4}	T_{BW}	T_{JT}	T_B	T_{B_4}	T_{BW}	T_{JT}	T_B	T_{B_4}	T_{BW}	T_{JT}	T_B
(i)	.3853	.3923	.3687	.2353	.2815	.2822	.2716	.1633	.1379	.1249	.1330	.0852
(ii)	.6633	.6614	.6413	.4714	.4902	.4892	.4736	.3186	.1936	.1788	.1913	.0212
(iii)	.9257	.9019	.9738	.9827	.8456	.8321	.8557	.9102	.3123	.3221	.3630	.3554
(iv)	1	1	1	1	1	1	.9999	.9993	.8001	.7601	.8135	.6507

					Double-exponential Samples							
$\sigma-$ pattern	(i)				(ii)				(iii)			
$\mu-$ pattern	T_{B_4}	T_{BW}	T_{JT}	T_B	T_{B_4}	T_{BW}	T_{JT}	T_B	T_{B_4}	T_{BW}	T_{JT}	T_B
(i)	.2670	.2751	.3233	.1941	.2040	.2077	.2448	.1408	.1185	.1097	.1275	.0758
(ii)	.4451	.4559	.5408	.3686	.3311	.3378	.4151	.2557	.1545	.1458	.1876	.1123
(iii)	.7098	.8106	.8532	.9004	.6271	.6380	.7273	.7853	.2522	.2419	.3449	.3272
(iv)	.9978	.9965	.9999	.9966	.9785	.9737	.9943	.9719	.5854	.5654	.7546	.5867

					Logistic Samples							
$\sigma-$ pattern	(i)				(ii)				(iii)			
$\mu-$ pattern	T_{B_4}	T_{BW}	T_{JT}	T_B	T_{B_4}	T_{BW}	T_{JT}	T_B	T_{B_4}	T_{BW}	T_{JT}	T_B
(i)	.1943	.2012	.1957	.0649	.1545	.1587	.1541	.0850	.1018	.0960	.0943	.0649
(ii)	.3163	.3205	.3250	.0785	.2381	.2383	.2442	.1405	.1272	.1172	.1252	.0785
(iii)	.6107	.6102	.6245	.1544	.4571	.4583	.4703	.4856	.1895	.1773	.2048	.1674
(iv)	.9806	.9792	.9864	.3082	.9127	.9058	.9244	.8377	.4215	.3960	.4703	.3082

					Exponential Samples							
$\sigma-$ pattern	(i)				(ii)				(iii)			
$\mu-$ pattern	T_{B_4}	T_{BW}	T_{JT}	T_B	T_{B_4}	T_{BW}	T_{JT}	T_B	T_{B_4}	T_{BW}	T_{JT}	T_B
(i)	.3888	.3881	.3719	.4539	.4873	.4832	.4371	.3059	.6321	.5855	.4843	.1416
(ii)	.6640	.6620	.6233	.7426	.7876	.7920	.7255	.6393	.9110	.8899	.7853	.4820
(iii)	.9689	.9694	.9455	.9958	.9951	.9955	.9864	.9927	.9990	.9991	.9985	.9920
(iv)	1.0000	1.0000	1.0000	1.0000	.9127	1.0000	1.0000	1.0000	1.0000	1.0000	1.0000	1.0000

					Cauchy Samples							
$\sigma-$pattern	(i)				(ii)				(iii)			
$\mu-$ pattern	T_{B_4}	T_{BW}	T_{JT}	T_B	T_{B_4}	T_{BW}	T_{JT}	T_B	T_{B_4}	T_{BW}	T_{JT}	T_B
(i)	.0710	.0602	.1847	.0973	.0637	.0516	.1498	.0785	.0495	.0365	.0943	.0564
(ii)	.0950	.0858	.2953	.1707	.0795	.0692	.2283	.1255	.0568	.0444	.1220	.0720
(iii)	.1536	.1515	.5419	.5150	.1261	.4583	.1202	.4167	.0740	.0610	.1998	.1646
(iv)	.3189	.3389	.9237	.7868	.2548	.2670	.8271	.6632	.1250	.1209	.4486	.2880

Note: $\sigma-$ pattern: (i) $\sigma_i = 1, i = 1, 2, 3, 4$; (ii) $\sigma_1 = 1, \sigma_2 = 1.2, \sigma_3 = 1.3, \sigma_4 = 1.5$; (iii) $\sigma_1 = 1, \sigma_2 = 2, \sigma_3 = 3, \sigma_4 = 4$.
$\mu-$ pattern: (i) $\mu_1 = 0, \mu_2 = 0.2, \mu_3 = 0.4, \mu_4 = 0.6$; (ii) $\mu_1 = 0, \mu_2 = 0.3, \mu_3 = 0.6, \mu_4 = 0.9$; (iii) $\mu_1 = 0, \mu_2 = 1, \mu_3 = 1.3, \mu_4 = 2$; (iv) $\mu_1 = 0, \mu_2 = 1, \mu_3 = 2, \mu_4 = 3$.

Table 4. Empirical power for 5% size tests for different distributions for $k = 4$, sample-sizes: $(10,15,20,25)$.

	Normal Samples											
$\sigma-$ pattern	(i)				(ii)				(iii)			
$\mu-$ pattern	T_{B_4}	T_{BW}	T_{JT}	T_B	T_{B_4}	T_{BW}	T_{JT}	T_B	T_{B_4}	T_{BW}	T_{JT}	T_B
(i)	.5188	.5078	.5113	.3639	.3419	.4128	.3658	.2469	.0789	.1802	.1339	.1022
(ii)	.8151	.7995	.8089	.6834	.6227	.6912	.6230	.4778	.1496	.2844	.2200	.1562
(iii)	.9955	.9930	.9947	.9979	.9646	.9753	.9481	.9658	.3655	.5707	.4451	.4451
(iv)	1	1	1	1	1	1	1	1	.9350	.9774	.9214	.8233
	Double-exponential Samples											
$\sigma-$ pattern	(i)				(ii)				(iii)			
$\mu-$ pattern	T_{B_4}	T_{BW}	T_{JT}	T_B	T_{B_4}	T_{BW}	T_{JT}	T_B	T_{B_4}	T_{BW}	T_{JT}	T_B
(i)	.3423	.3532	.4359	.2908	.2167	.2830	.3161	.1990	.0580	.1343	.1394	.0930
(ii)	.5789	.5859	.7114	.5574	.3946	.4746	.5380	.3919	.0874	.2044	.2139	.1410
(iii)	.9103	.8984	.9720	.9739	.7677	.8194	.8868	.8881	.2095	3839	.4192	.4027
(iv)	1	.9995	1	1	.9992	.9990	.9999	.9986	.6765	.8260	.8790	.7629
	Logistic Samples											
$\sigma-$ pattern	(i)				(ii)				(iii)			
$\mu-$ pattern	T_{B_4}	T_{BW}	T_{JT}	T_B	T_{B_4}	T_{BW}	T_{JT}	T_B	T_{B_4}	T_{BW}	T_{JT}	T_B
(i)	.2468	.2472	.2619	.1542	.1525	.2038	.1904	.0772	.0399	.1089	.1004	.0608
(ii)	.4156	.4158	.4439	.2973	.2698	.3337	.3183	.1273	.0616	.1506	.1335	.0731
(iii)	.7650	.7554	.8072	.8155	.5688	.6457	.6268	.4179	.1297	.2722	.2313	.1475
(iv)	.9990	.9974	.9997	.9967	.9813	.9867	.9889	.7910	.4589	.6496	.5773	.2757
	Exponential Samples											
$\sigma-$ pattern	(i)				(ii)				(iii)			
$\mu-$ pattern	T_{B_4}	T_{BW}	T_{JT}	T_B	T_{B_4}	T_{BW}	T_{JT}	T_B	T_{B_4}	T_{BW}	T_{JT}	T_B
(i)	.5263	.5101	.5114	.5771	.6256	.6916	.6117	.4146	.7794	.9006	.6880	.2866
(ii)	.8211	.8074	.7998	.8531	.9205	.9460	.9017	.7716	.9895	.9986	.9496	.7392
(iii)	.9963	.9975	.9930	.9986	.9999	.9999	.9996	.9980	1.0000	1.0000	1.0000	.9994
(iv)	1.0000	1.0000	1.0000	1.0000	1.0000	1.0000	1.0000	1.0000	1.0000	1.0000	1.0000	1.0000
	Cauchy Samples											
$\sigma-$pattern	(i)				(ii)				(iii)			
$\mu-$ pattern	T_{B_4}	T_{BW}	T_{JT}	T_B	T_{B_4}	T_{BW}	T_{JT}	T_B	T_{B_4}	T_{BW}	T_{JT}	T_B
(i)	.0811	.0604	.2376	.0973	.0591	.0559	.1809	.0790	.0260	.0386	.0955	.0564
(ii)	.0984	.0949	.4004	.1707	.0704	.0759	.2909	.1253	.0293	.0465	.1348	.0720
(iii)	.1451	.1630	.7256	.5150	.1005	.1253	.5581	.3856	.0363	.0649	.3182	.1646
(iv)	.2922	.3498	.9882	.9495	.2094	.2778	.9449	.6607	.0649	.1266	.5531	.2880

Note: $\sigma-$ pattern: (i): $\sigma_i = 1, i = 1, 2, 3, 4$; (ii): $\sigma_1 = 1, \sigma_2 = 1.2, \sigma_3 = 1.3, \sigma_4 = 1.5$; (iii): $\sigma_1 = 1, \sigma_2 = 2, \sigma_3 = 3, \sigma_4 = 4$.
$\mu-$ pattern: (i): $\mu_1 = 0, \mu_2 = 0.2, \mu_3 = 0.4, \mu_4 = 0.6$; (ii): $\mu_1 = 0, \mu_2 = 0.3, \mu_3 = 0.6, \mu_4 = 0.9$; (iii): $\mu_1 = 0, \mu_2 = 1, \mu_3 = 1.3, \mu_4 = 2$; (iv): $\mu_1 = 0, \mu_2 = 1, \mu_3 = 2, \mu_4 = 3$.

3. A Simulation Study

We use Monte Carlo simulation in order to study the α robustness and the power of the tests described for normal and non-normal data. The distributions considered in this study are normal, double-exponential, logistic, exponential and Cauchy. Sample sizes considered were $10, 10, 10, 10$ and $10, 15, 20, 25$. The levels of the tests and powers are based on 100,000 replications. The location parameters for the null distributions are chosen to be zero.

The sizes are displayed in Tables 1-2 for $\alpha = .10, .05$ and $.01$ for three variance patterns, (i): $\sigma_i = 1, i = 1, 2, 3, 4$; (ii): $\sigma_1 = 1, \sigma_2 = 1.2, \sigma_3 = 1.3, \sigma_4 = 1.4$; (iii): $\sigma_1 = 1, \sigma_2 = 2, \sigma_3 = 3, \sigma_4 = 4$. The power study considers four patterns for location parameters, namely (i): $\mu_1 = 0, \mu_2 = 0.2, \mu_3 = 0.4, \mu_4 = 0.6$; (ii): $\mu_1 = 0, \mu_2 = 0.3, \mu_3 = 0.6, \mu_4 = 0.9$; (iii): $\mu_1 = 0, \mu_2 = 1, \mu_3 = 1.3, \mu_4 = 2$; (iv): $\mu_1 = 0, \mu_2 = 1, \mu_3 = 2, \mu_4 = 3$, for each of different populations and each of the three scale patterns. These are displayed in Tables 3 and 4 for the two sets of sample sizes.

4. Results and Discussion

4.1. *Comparison of level of the tests*

Based on Tables 1 and 2 we observe that:

- In case of equal sample sizes for both equal and unequal variances, the statistics T_{B_4},, T_{BW}, T_{JT} and T_B due to Bathke are quite α-robust, for almost all distributions except in some cases involving Cauchy distribution.
- Barlow *et al.* test as adapted for unequal variances does not satisfy the level robustness for Cauchy sample when scales are unequal. But the test due to Bathke is $\alpha-$robust in case of Cauchy distribution. It, however, breaks down in case of exponential distribution for unequal variances with respect to its size.
- It is also observed that for unequal sample sizes, almost all the statistics are quite $\alpha-$robust for almost all distributions considered in this study; the Barlow *et al.* statistic B seems to be conservative for Cauchy distribution for unequal sample sizes.

4.2. *Comparison of the power of the tests*

Power of the tests are shown in Tables 3-4 for 5% size of the tests. From these figures we observe that:

- When the underlying distribution is normal and the variances are equal then for both the cases of equal or unequal sample sizes, the B and BW tests are more powerful than other tests. For population with unequal variances the BW test is powerful for both the cases of equal and unequal sample sizes.
- For populations like double exponential, logistic and Cauchy which have long-tails, the statistic T_{JT} can be highly recommended without second thought for both the cases of equal and unequal sample sizes having equal and unequal variances. However, the Barlow *et al.* statistic may be preferred for some particular cases of logistic population..
- The power of Bathke's test strongly depends on the underlying distribution. It is recommended for exponential distribution for both the cases of equal and unequal sample sizes having equal variances. However, the Barlow *et al.* statistic B can be recommended for the cases of equal sample sizes with unequal variances.

5. Conclusions

From the above study of $\alpha-$robustness and power of the tests we can conclude that

1. Bathke's statistic turns out to be more powerful for exponential distribution for both the cases of equal and unequal sample sizes with equal variances.
2. JT statistic is recommended for double exponential, logistic and Cauchy populations in cases of both equal and unequal sample sizes with equal and unequal scales.
3. In case of normal population with equal and unequal sample sizes having equal variances, the Barlow et al. test is the only candidate with high power while its adaptation for unequal variances may be considered for the populations with unequal variances.
4. This study suggests that Bathke's test may be more powerful than some competing tests for asymmetric populations, however, JT test may still be quite appropriate for symmetric populations.

References

1. R. E. Barlow, D. J. Bartholomew, J. M. Bremner and H. D. Brunk, *Statistical Inference under Order Restrictions* (John Wiley and Sons, New York, 1972).
2. D. J. Bartholomew, *Biometrika* **46**, 36 (1959).

3. D. J. Bartholomew, *Biometrika* **46**, 328 (1959).
4. D. J. Bartholomew, *Biometrika* **48**, 325 (1961).
5. A. C. Bathke, *Metrika* **69**, 17 (2009).
6. M. L. Berenson, *Biometrika* **47**, 265 (1982).
7. M. L. Berenson and S. T. Gross, *Communication in Statistics - Simulation and Computation* **B10**, 405 (1981).
8. H. Büning, *OR Spektrum* **16**, 33 (1994).
9. H. Büning, *Communications in Statistics - Theory and Methods* **A25**, 1569 (1996).
10. H. Büning, *Journal of Applied Statistics* **24**, 319 (1997).
11. H. Büning, *Journal of Applied Statistics* **26**, 541 (1999).
12. H. Büning, and W. Kössler, *Journal of Statistical Computation and Simulation* **55**, 319 (1996).
13. J. D. Gibbons and S. Chakraborti, *Nonparametric Statistical Inference, 5th edition* (CRC Press, Florida, USA, 2010).
14. B. Gogoi and S. Kakoty, *Jour. Ind. Stat. Assoc.* **25**, 37 (1987).
15. T. P. Hettmansperger and R. M. Norton, *Journal of the American Statistical Association* **397**, 292 (1987).
16. K. Jogdeo, *Annals of Mathematical Statistics* **37**, 1697 (1966).
17. A. R. Jonckheere, *Biometrika* **41**, 133 (1954).
18. M. L. Puri, *Communications in Pure and Applied Mathematics* **18**, 51 (1965).
19. F. L. Ramsey, *Journal of the American Statistical Association* **66**, 149 (1971).
20. T. J. Terpstra, *Indagationes Mathematicae* **14**, 327 (1952).
21. B. L. Welch, *Biometrika* **29**, 350 (1937).

LOCAL SEARCH FOR IDENTIFYING COMMUNITIES IN LARGE RANDOM GRAPHS

NARSINGH DEO[*] and MAHADEVAN VASUDEVAN[†]

Dept. of EECS, University of Central Florida, Orlando, FL
[*]E-mail: deo@eecs.ucf.edu, [†]E-mail: maha@knights.ucf.edu

A community may be defined informally as a locally-dense subgraph, of a significant size, in a large, globally-sparse graph. Communities do not exist in the classical Erdös-Rényi random graph, but they do exist in graphs representing the Internet, the World Wide Web (WWW), and numerous social and biological networks. At least two different questions may be posed about the community structure in large graphs: (i) Given a graph, identify or extract all (i.e., sets of nodes that constitute) communities; and (ii) Given a node (or a small subset of seed nodes) in the graph identify the best community to which the given node(s) belong, if there exists such a community. Several algorithms have been proposed to solve the former problem, known as Community Discovery. The latter problem, known as Community Identification, has also been studied, but to a much smaller extent. Both the problem of community discovery and community identification have been shown to be NP-complete. In this paper, we will first discuss more precise graph-theoretic definitions of communities and their presence in the context of other non-Erdös-Rényi properties and then propose a taxonomy—based on underlying solution strategies—of existing algorithms for community structures. We will also present a fast, approximate, local-search heuristic for identifying the community to which a given node may belong—using only the neighborhood information.

1. Introduction

Complex network refers to any large, dynamic, random graph that corresponds to the interactions among entities in real-world complex systems. Each *node* of the network corresponds to an individual object of the system and the *edge* symbolizes the interaction between the individual entities (see Strogatz[51]). These large random graphs exist prominently across disciplines such as sociology, biology, computer science, linguistics and mathematics. Examples of real-world complex networks include the Internet, the World Wide Web (WWW), protein-interaction networks, human metabolic networks, ecological networks and railroad networks (Boccaletti *et al.*,[6] Newman[34]). These networks exhibit randomness significantly different from that of the *classical Erdös-Rényi random graphs* (Erdös and Rényi[17]). Therefore, over the past 10 to 15 years engineers and scientists have been trying to study the properties of such complex networks.

Despite the large size of the complex networks, certain network properties and behavior have been studied in detail while few others still remain in exploratory levels. Their properties have been studied mainly at two extreme levels: (*i*) *microscopic* - properties at node level (degree distribution as in Watts and Strogatz[55] and clustering coefficient as in Cami and Deo[8]), and (*ii*) *macroscopic* - global properties such as the network distance (small-world effect as in Albert *et al.*[3] and Caci *et al.*[7]). In general, complex networks follow *power-law* degree distribution (Albert *et al.*[3]), and high clustering coefficient values (greater than 0.3) (Newman[35]) of nodes in the network. The power-law coefficient, average distance and clustering coefficient have been empirically identified for more than 700 real-world complex networks (Onnela *et al.*[37]). It is these statistically significant properties that make them stand out as a new science of networks.

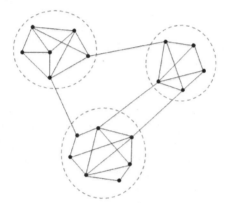

Fig. 1. Communities in a graph. The subgraphs marked
by dashed circles correspond to communities.

However, recent research has focused on the study of closely-knit subgraphs in a relatively sparse neighborhood referred as *Communities*. Modules, motifs, and clusters are other terminologies referring to such dense subgraphs. Figure 1 shows an example of a graph with three communities. They lie intermediate between the microscopic and macroscopic properties and hence they are referred as a *mesoscopic* property. The recent growth of the internet and online social networks, and the computational advancements has triggered interest in communities on a large scale. Techniques to define, identify, extract and detect communities in networks are classified as *community detection* algorithms.

Global community detection i.e., to extract all the communities in the given network, is tedious especially in the case of complex networks. It is shown by Fortunato[21] that detecting communities in a graph is NP-complete. It necessitates the exploration of the entire graph. Whereas, identifying communities from a given seed node using only its local neighborhood information is relatively easier and has wide range of real-world applications (see Agarwal *et al.,*[1] Shniederman[50] and Strogatz[51]). But identifying an induced connected subgraph with a specific property in a given graph is also NP-complete (Garey and Johnson[22]). In this paper, we will first discuss more precise graph-theoretic definitions of communities and their presence in the context of other non-Erdös-Rényi properties and then propose a taxonomy—based on underlying solution strategies—of existing algorithms for community structures. We will also present a fast, approximate, local-search heuristic for identifying the community to which a given node may belong--using only the neighborhood information.

2. Defining a Community

Informally, a community maybe defined as a locally-dense subgraph in a large globally-sparse graph. An accepted mathematical definition for a community in graphs is yet to hit the literature, because it is difficult to exemplify its relatively dense nature (see Fortunato[21]). Existing definitions of a community are conceptual and depends mainly on the topology of the underlying network. The relation every node shares with its neighbors, both within and outside the subgraph forms the essence in those definitions. Some of the definitions are also constructive, i.e., the result of algorithmic steps (see Scanlon and Deo[45]). We classify the community definitions in the literature into three main categories – (i) *diameter*[a], (ii) *degree*[b], and (iii) *alliance*. Earlier notion of a community even focused on its equivalence to a *clique* (see Luce and Perry[29]). But expecting every member to be connected to every other member within a subgraph of a sparse random network is stringent. Table 1 gives a brief listing of the various definitions for a community from the literature. It is evident that a community needs to be defined in terms of the degree of the nodes. Especially, in the case of a large sparse graph, a dense subgraph should have nodes with higher *average degree* than those randomly present. We propose a new definition for a community based on this notion.

Our proposed definition of a community is as follows: Given a graph $G(V, E)$, the average degree of an induced subgraph $g(V', E')$ is defined as the ratio of twice the number of edges in the subgraph to the total number of nodes in g. Average degree of the subgraph g is given by

$$\overline{d}_{V'} = \frac{2*|E'|}{|V'|}.$$

Note that the sum of the degrees of all nodes in a graph is twice the number of edges in the graph (see Deo[13]). An example graph with its subgraph and average degree is shown in Figure 2.

---

[a] The *diameter* of a graph is defined as the largest distance between two nodes in the graph.
[b] The number of edges incident on a node is called the *degree* of the node.

Table 1. Existing definitions of a community.

Definitions of a Community		
Clique	Every vertex is adjacent to every other vertex	(Clique) - Luce and Perry[29]
Diameter	diameter of $g \leq d$	(d-clique) - Alba[2]; (d-clan & d-club) - Mokken[32]
Degree	Every vertex is adjacent to at least δ other vertices in the subgraph [internal degree of every vertex $\geq \delta$]	(δ-core) - Seidman[48]; (δ-plex) - Seidman and Foster[49]
	Sum of internal degrees > Sum of external degree	[Weak-community] – Radicchhi *et al.*[43]
	Sum of internal degrees > the number of edges the subgraph shares with other communities	[Weak-community] – Hu *et al.*[24]
Alliance	Internal degree of each vertex > External degree of the vertex	[Strong alliance] - Radicchhi *et al.*[43]
	Internal degree of every vertex is greater than the number of edges the vertex shares with other communities	[Strong alliance] - Hu *et al.*[24]
	Internal degree of each vertex $\geq$ External degree of the vertex	[Defensive alliance] – Flake *et al.*[18]
	Set of nodes such that each of its proper subsets has more edges to its complement within the set than outside	[LS-set] – Luccio and Sammi[28]

The subgraph $g(V', E')$ forms a community if the average degree of g ($\bar{d}_{V'}$) is greater than the average degree of any other subset of vertices within the neighborhood of V'. Since complex networks are large, we restrict ourselves to considering only the set of vertices within a neighborhood at a distance k. Therefore, if $\Gamma(k, u)$ denotes the set of vertices within a neighborhood k from vertex u, then $V' \subseteq \Gamma(k, u)$ forms a community if

$$\forall V'' \subseteq \Gamma(k, u), \qquad \bar{d}_{V''} < \bar{d}_{V'}$$

Since, in a community identification algorithm (discussed in the next section) we begin with a seed vertex $u \in V$, defining a community by considering the vertex within k-neighborhood is a realistic measure. The algorithm we propose (in section 4) is based on this definition and it effectively identifies communities in large networks. Global community discovery can also utilize this definition and look out for subgraphs with maximum average degree.

Fig. 2. Average degree of g (shown by dotted line) is 3.2 $[\frac{2 \times 8}{5}]$.

3. Community Detection Algorithms

Algorithms dealing with community detection in complex networks address one of the following two questions: (*i*) given a network, can we explore and extract subsets of nodes that form a community? (*ii*) given a network and a seed node, can we identify the best community that the given seed belongs to, if there exists one? The former problem, known as *Community Discovery*, has been studied extensively in the literature and a number of approximate algorithms has been proposed (see Clauset *et al.*,[12] Flake *et al.*,[19] Girvan and Newman,[23] Newman[33] and Rives and Galitsky[43]). The community discovery algorithms further branch down based on the following: *Community Detection, Community Partition* and *Overlapping Communities (Cover)*. The latter, known as *Community Identification*, has also been studied in the literature, but to a smaller extent. Community identification can be formally described as follows: Given a large sparse graph G (V, E) and a seed vertex $u \in V$, does there exist a community that u belongs to? If yes, return the induced subgraph. Community identification algorithms are very sparse in the literature and are mainly greedy heuristics (see Bagrow,[4] Chen *et al.*,[9] Clauset,[11] Luo *et al.*[30] and Seidman[47]).

Techniques to discover communities in complex networks are broadly discussed by a few authors (see Fortunato,[21] Porter *et al.*,[41,42] Schaeffer,[46] and Vasudevan *et al.*[53]), but none of them elucidate community identification algorithms in detail. The taxonomy in Fig. 3 gives a comprehensive classification of the existing algorithms to detect communities. The algorithm we propose belongs to the community identification category and is explained in the following section.

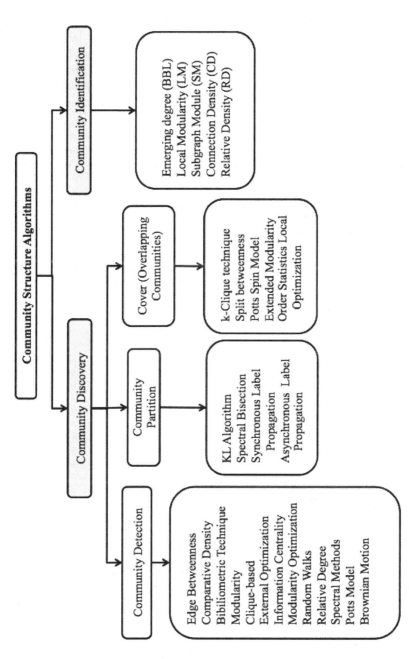

Fig. 3. Taxonomy of Community Detection Algorithms.

4. Proposed Community Identification Algorithm

Our community definition clearly indicates that we are interested in identifying subsets of vertices with maximal average degree in a given graph. Given a large graph and a seed node u we identify a set of vertices that u belongs to satisfying the above criteria. Since we explore only a restricted neighborhood of u, there is one other parameter k, required as input to the algorithm. In other words, the parameter k specifies the distance till which we explore the given graph from the seed. Typically, the value of k is a positive integer greater than two. When the value of k is 1 or 2, we have just the seed node and/or few of its neighbors (neighbors of neighbors), which is insufficient to identify a community.

ALGORITHM 1:

Procedure *find_Community* ( )
{
$\quad C \leftarrow BFS(G, u, k)$

$\quad d_C \leftarrow Compute\_Avg\_Deg\ (C)$

$\quad$**do**
$\quad${

$\qquad H \leftarrow C$

$\qquad d_H \leftarrow d_C$

$\qquad d_{min} \leftarrow \min_{v \in V_C} d_v$

$\qquad V' \leftarrow \{v \in V_C \mid d_v > d_{min}\}$

$\qquad V_C \leftarrow V'$

$\qquad E_C \leftarrow \{(u,v) \mid u,v \in V_C\} \cap E_C$

$\qquad d_C \leftarrow Compute\_Avg\_Deg\ (C)$

$\quad$**}while** $(d_C > d_H)$

$\quad C \leftarrow H$

}

As a first step of the algorithm we perform a *breadth-first search* (see Deo[13]) on the input graph starting at node u. Once we obtain the subset of vertices $\Gamma(k,u)$ by the breadth-first technique, our algorithm iteratively removes nodes that do not belong to the community. Let C denote this initial set of vertices $\Gamma(k,u)$. The average degree of the subgraph $[\bar{d}_C]$ is calculated. Now, we remove the nodes with minimum degree and calculate the new average degree $[\bar{d}_H]$. We compare the two average degree values and if $\bar{d}_H$ is greater than $\bar{d}_C$

then we store the new set of vertices in C. It is to be noted that the degree of the nodes in C needs to be updated on the removal of minimum degree nodes. Again, with the new set of vertices, we remove the nodes with the least degree and re-compute the average degree $\bar{d}_H$. The iterative removal and average degree computation continues till $\bar{d}_H$ is greater than $\bar{d}_C$. If $\bar{d}_H$ becomes less than $\bar{d}_C$, we stop the iteration and return the set of nodes C as the community. If a given seed does not belong to any community, it will be removed in one of the iterations of the algorithm [since the node would not contribute to a higher average density of the resulting set of nodes]. A detailed description of the notation and steps involved in the algorithm are given by Algorithm 1.

The time complexity of the algorithm is $O(n\bar{d})$, where n denotes the number of nodes in the k-neighborhood [$n = \Gamma(k, u)$] and $\bar{d}$ denotes the average degree of the nodes of the subgraph. For each of the nodes (n) in the resulting community, the algorithm needs to check the degree of the adjacent nodes ($n\bar{d}$).

5. Experimental Results

Synthetic and real-world graphs have been used as benchmarks to test the effectiveness (speed and accuracy) of community detection algorithms. Since the community identification problem is NP-hard, any practical algorithm is expected to produce only approximate solution. Thus the goodness of the algorithm must be measured by (i) the quality of the community produced and (ii) the execution time. Classical random graphs do not exhibit communities, since there is no preferential attachment among nodes to form denser subgraphs [8]. Several random-graph generators take into account the dynamic addition of nodes and edges and produce graphs similar to real-world complex networks (see Chung et al.,[10] Deo and Cami,[14] Flaxman et al.,[20] Newman,[34] Pollner et al.,[40] and Vázquez[54]). Similarly, numerous examples of complex networks exist in real world such as the web graph (Dourisboure et al.[16]), protein-interaction networks (Luo et al.[31]) and the Internet (Barbási[5]). The execution results of our algorithm on these synthetic graphs and some real-world graphs are discussed below.

5.1 Synthetic graphs

It is essential that the algorithms are tested on graphs with known community structures to see if they recognize the already known community structure. Newman and Girvan[36] described one such graph with 128 nodes divided equally into four communities [GN graph]. We generated GN graphs with 256 [GN256] and 512 [GN512] nodes and tested all five existing algorithms (mentioned in Figure 3) with our algorithm. A comparison of the execution times of the algorithms is shown in Figure 4. All the algorithms were given the same seed and their execution time to identify the community is calculated in milliseconds. The proposed algorithm (AD) clearly has better performance in terms of speed compared to other algorithms. Of these we eliminated BBL algorithm from further consideration because it did not identify known communities in GN graphs.

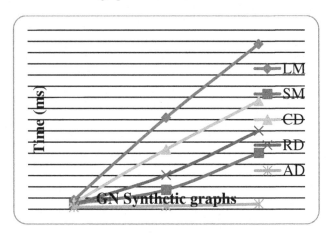

Fig. 4. Run time comparison of Community identification algorithms.

In order to compare the accuracy, we plotted the results from our algorithm along with the two faster algorithms (refer Figure 4) — Luo *et al.*'s[30] algorithm using *Subgraph Modularity* (*SM*) and Schaeffer's[47] algorithm based on *Relative Density* (*RD*). The vertices of the GN graph were divided into four communities of equal size with each vertex consisting of 16 adjacent nodes ($d_i = 16$). The number of inter-community edges was varied from one to nine ($1 \leq d_{out} \leq 9$). Therefore, a total of nine different GN graphs were generated as test cases. The algorithms were tested using the same seed in each of the test cases and *Jaccard's index* $J(T,C)$ was used to calculate the percentage of nodes correctly identified. Here, T denotes the target community vertices for a given seed vertex and C denotes the set of vertices identified for a seed by a community identification algorithm.

All three algorithms identify the community nodes with no false or true positives until the number of inter-community edges per vertex is less than or equal to 6 [$d_{out} \leq 6$]. But for $d_{out} \geq 7$, *SM* and *RD* algorithms fail to identify all the nodes in the community a given seed belongs to, whereas our algorithm has $J(T,C)$ equal to one (Figure 5).

Another model for generating a controlled test graph with communities was proposed by Lancichinetti *et al.*[26,27]. These graphs are referred as LFR (Lancichinetti, Fortunato and Radicchi) graphs and consist of communities of different sizes unlike GN graphs. LFR graphs are better in testing a community algorithm because the communities are of variable sizes and the average degree also varies. We compared the results of our algorithm with Luo *et al.*'s[30] algorithm (SM) and the Jaccard index values are shown in Figure 6. The RD algorithm did not effectively identify communities in LFR graphs, so we have ignored it for comparison.

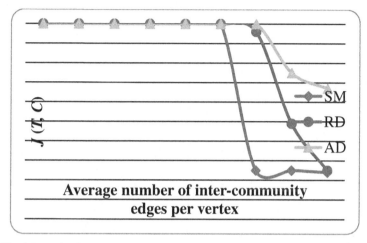

Fig. 5. Jaccard index comparing results of three community identification algorithms on communities with varying inter-community edges.

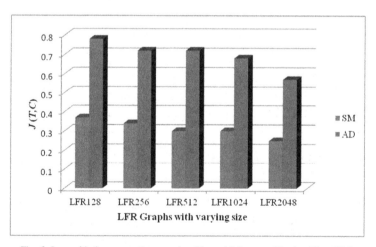

Fig. 6. Jaccard index comparing our algorithm with Luo *et al.*'s algorithm (SM) on LFR graphs with varying size. LFR*x* refers to graph with *x* nodes.

Eventually, we are interested in the same set of nodes that contribute to the community, irrespective of the node from which we begin the identification. Figure 7 depicts an interesting comparison for seed nodes from the same community but of varying degrees. Nodes with relatively lower degree in the community do not lead to identifying the entire expected set of nodes in the community and hence the lower value of Jaccard index. The consistency of our algorithm irrespective of the initial seed is evident from the Jaccard index range (0.65 to 1.0).

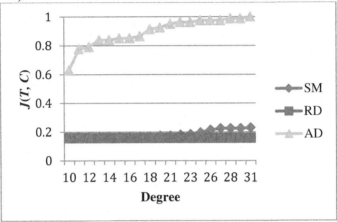

Fig. 7. Jaccard index comparing results from three community identification algorithms for input seed with different degree.

5.2 Real-world networks

Table 2. List of complex networks (n – number of nodes, m – number of edges, $\bar{d}$ – average degree of the graph and N_c – number of communities)

Network	Type	n	m	$\bar{d}$	N_c
Zachary Karate Club	Social	34	78	4.6	2
Bottlenose dolphins	Social	62	159	5.1	2
Brain Functions	Biological	90	337	7.5	5
Books on American Politics	Business	105	441	8.4	2
NCAA Football (2011)	Social	120	674	5.6	11
Jazz Musicians	Social	198	2742	27.7	2

Numerous examples of real-world networks have been used as benchmarks to test community detection algorithms in the literature (see Albert *et al.*,[3] Dorogovtsev and Mendes,[15] Jeong *et al.*,[25] Pellegrini *et al.*,[38] Pimm[39] and Zachary[56]). Some of the real-world benchmarks in the literature are not large networks (less than 500 nodes), but are still considered because of their well-defined community structure. For example, Zachary's[56] karate club network is the most cited real-world social network (in community detection literature), but it is a very small network (34 nodes). We tested our algorithm on several known real-world graphs. Table 2 gives a list of real-world complex networks that we used as input to our algorithm to identify communities.

6. Conclusion

We have proposed a fast and efficient algorithm for identifying communities in real-world complex networks. The experimental result section is evident of the ability of our algorithm to identify statistically significant communities in any given network. We have also proposed a definition for a community in complex network based on the average degree of an induced subgraph, which betters the existing definitions. The concept of locally-dense subgraph in a given globally-sparse graph is aptly captured in our definition of a community. A comprehensive taxonomy classifying the existing community detection algorithms was also presented. Community identification has a wide range of applications as mentioned before. Algorithms to identify dense subgraphs can be used in movie and music recommender systems to predict and recommend movies (and music) of similar genre. Modeling such recommender systems as graphs and identifying communities in such graphs would serve as interesting directions for future research.

Acknowledgement

The authors wish to thank the anonymous referee for making a valuable suggestion, which led to the rewriting of Section 5.1 and inclusion of Figure 4. The paper has been improved as a consequence.

References

1. N. Agarwal, H. Liu, L. Tang and P. Yu, *Social Network Analysis And Mining* **2**, 139 (2012).
2. R. D. Alba, *The Journal Of Mathematical Sociology* **3**, 113 (1973).
3. R. Albert, H. Jeong and A. Barabasi, *Nature* **401**, 130 (1999).
4. J. P. Bagrow, *Journal of Statistical Mechanics: Theory And Experiment* **2008**, 001 (2008).
5. A. Barabási, *Science* **325**, 412 (2009).
6. S. Boccaletti, V. Latora, Y. Moreno, M. Chavez and D. U. Hwang, *Physics Reports* **424**, 175 (2006).
7. B. Caci, M. Cardaci and M. Tabacchi, *Social Network Analysis And Mining* **2**, 163 (2012).
8. A. Cami and N. Deo, *Networks* **51**, 211 (2007).
9. J. Chen, O. Zaiane and R. Goebel, *Proceedings of the International Conference on Advances in Social Network Analysis and Mining (ASONAM)*, 237 (2009).
10. F. Chung, L. Lu, T. G. Dewey and D. J. Galas, *Journal of Computational Biology* **10**, 677 (2003).

11. A. Clauset, *Phys. Rev.* **E 72,** 026132 (2005).
12. A. Clauset, M. E. J. Newman and C. Moore, *Phys. Rev.* **E 70,** 066111 (2004).
13. N. Deo, *Graph Theory with Applications to Engineering and Computer Science* (Prentice-Hall Inc, Upper Saddle River, NJ, USA, 1974).
14. N. Deo and A. Cami, *Info. Processing Lett.* **102,** 156 (2007).
15. S. N. Dorogovtsev And J. F. F. Mendes, *Evolution Of Networks: From Biological Nets to the Internet and WWW* (Oxford University Press, London, 2003).
16. Y. Dourisboure, F. Geraci And M. Pellegrini, *Proc. of the 16th Intl. Conf. on World Wide Web*, ACM, Banff, Alberta, Canada, 461 (2007).
17. P. Erdös And A. Rényi, *Publ. Math. Debrecen* **6,** 290 (1959).
18. G. W. Flake, S. Lawrence and C. L. Giles, *Proceedings of the Sixth ACM SIGKDD International Conference on Knowledge Discovery and Data Mining* ACM, New York, United States, 150 (2000).
19. G. W. Flake, S. Lawrence, C. Lee Giles and F. M. Coetzee, *Computer* **35,** 66 (2002).
20. A. D. Flaxman, A. M. Fricze and J. Vera, *Internet Math* **3,** 187 (2006).
21. S. Fortunato, *Physics Report* **486,** 75 (2010).
22. M. R. Garey and D. S. Johnson, *Computers and Intractability: A Guide to the Theory of NP-Completeness* (Freeman, New York, USA, 1979).
23. M. Girvan and M. E. J. Newman, *Proc. Of Natl. Acad. Sci. USA* **99,** 7821 (2002).
24. Y. Hu, H. Chen, P. Zhang, M. Li, Z. Di and Y. Fan, *Phys. Rev.* **E 78,** 026121 (2008).
25. H. Jeong, S. P. Mason, A. L. Barabasi and Z. N. Oltvai, *Nature* **411,** (2001).
26. A. Lancichinetti, S. Fortunato and F. Radicchi, *Phys. Rev.* **E 78,** 046110 (2008).
27. A. Lancichinetti, F. Radicchi, J. J. Ramasco and S. Fortunato, *PLoS ONE* **6,** e18961 (2011).
28. F. Luccio and M. Sami, *IEEE Transactions on Circuit Theory* **16,** 184 (1969).
29. R. Luce and A. Perry, *Psychometrika* **14,** 95 (1949).
30. F. Luo, J. Z. Wang and E. Promislow, *Web Intelligence and Agent Systems* **6,** 387 (2008).
31. F. Luo, Y. Yang, C.-F. Chen, R. Chang, J. Zhou and R. H. Scheuermann, *Bioinformatics* **23,** 207 (2007).
32. R. J. Mokken, *Quality and Quantity* **13,** 161 (1979).
33. M. E. J. Newman, *Eur. Phys. J.* **B 38,** 321 (2004).
34. M. E. J. Newman, *Siam Review* **45,** 167 (2003).
35. M. E. J. Newman, *Proc. of Natl. Acad. Sci. USA* **98,** 404 (2001).
36. M. E. J. Newman and M. Girvan, *Phys. Rev. E* 69, 026113 (2004).

37. J.-P. Onnela, D. Fenn, S. Reid, M. Porter, P. Mucha, M. Fricker and N. Jones, *A Taxonomy of Networks*, (2010); available at arXiv:1006.573.

38. M. Pellegrini, D. Haynor and J. M. Johnson, *Expert Rev. of Proteomics* **1**, 239 (2004).

39. S. L. Pimm, *Theoretical Population Biology* **16**, 144 (1979).

40. P. Pollner, G. Palla and T. Vicsek, *Eur. Phy. Lett.* **73**, 478 (2006).

41. M. A. Porter, J. Onnela and P. J. Mucha, *Notices of the AMS* **56**, 1082 (2009).

42. M. A. Porter, J. Onnela and P. J. Mucha, *Notices of the AMS* **56**, 1164 (2009).

43. F. Radicchi, C. Castellano, F. Cecconi, V. Loreto and D. Parisi, *Proc. Of Natl. Acad. Sci. USA*, **101**, 2658 (2004).

44. A. W. Rives and T. Galitski, *Proc. of Natl. Acad. Sci. US*, **100**, 1128 (2003).

45. J. M. Scanlon and N. Deo, *Congressus Numerantium* **190**, 183 (2008).

46. S. E. Schaeffer, *Computer Science Review* **1**, 27 (2007).

47. S. E. Schaeffer, In *Advances in Knowledge Discovery and Data Mining, (Eds.:* Tu Bao Ho, David Wai-Lok Cheung and Huan Liu), *Proc. of 9th Pacific-Asia Conf. on Knowledge Discovery and Data Mining, LNCS 3518,* (Springer, New York, 2005*).*

48. S. B. Seidman, *Social Networks* **5**, 269 (1983).

49. S. B. Seidman and B. L. Foster, *The Journal of Mathematical Sociology* **6**, 139 (1978).

50. B. Shneiderman, *IEEE Transactions on Visualization and Computer Graphics* **12**, 733 (2006).

51. Y. Song, Z. Zhuang, H. Li, Q. Zhao, J. Li, W.-C. Lee and C. L. Giles, *Proceedings of the 31st Annual International ACM SIGIR Conference on Research and Development In Information Retrieval*, ACM, Singapore, 515 (2008).

52. S. H. Strogatz, *Nature* **410**, 268 (2001).

53. M. Vasudevan, H. Balakrishnan and N. Deo, *Congressus Numerantium* **196**, 127 (2009).

54. A. Vázquez, *Phys. Rev. E* **67**, 056104 (2003).

55. D. J. Watts and S. H. Strogatz, *Nature* **393**, 440 (1998).

56. W. W. Zachary, *J. of Anthropological Research* **33**, 452 (1977).

DATA MINING APPROACHES TO MULTIVARIATE BIOMARKER DISCOVERY

DARIUS M. DZIUDA

Department of Mathematical Sciences,
Central Connecticut State University,
New Britain, USA
E-mail: dziudadad@ccsu.edu

Many biomarker discovery studies apply statistical and data mining approaches that are inappropriate for typical data sets generated by current high-throughput genomic and proteomic technologies. More sophisticated statistical methods should be used and combined with appropriate validation of their results as well as with methods allowing for linking biomarkers with existing or new biomedical knowledge. Although various supervised learning algorithms can be used to identify multivariate biomarkers, such biomarkers are often unstable and do not necessarily provide the insight into biological processes underlying phenotypic differences. In this paper, we will outline common misconceptions in biomarker discovery studies, and then focus on the methods and concepts leading to parsimonious multivariate biomarkers that are stable and biologically interpretable. First, we will describe the *Informative Set of Genes* defined as the set containing all of the information significant for class differentiation. Then, we will present the *Modified Bagging Schema* generating ensembles of classifiers that are used to vote for variables. Finally, we will show how to combine these methods to identify robust multivariate biomarkers with plausible biological interpretation.

1. Introduction

Biomarker discovery studies based on current high-throughput genomic and proteomic technologies analyze data sets with thousands of variables and much fewer biological samples. To successfully deal with such data, bioinformaticians or biomedical researchers have to be familiar with intricacies of multivariate analysis, and particularly with approaches allowing overcoming the *curse of dimensionality*.

One of the common misconceptions is to limit biomarker discovery studies to a univariate approach. Such an approach is a relic of the old "one-gene-at-a-time" paradigm, and is based on the unrealistic assumption that gene expression (or protein expression) variables are independent of one another. Although it is possible that expression of a single gene may be all that is necessary to efficiently separate phenotypic classes, such situations are exceptions rather than a rule. In most situations, the assumptions of no correlations and no interactions among genes are unwarranted. Furthermore, limiting biomarker discovery investigations to some number of top univariately significant variables may remove from consideration the variables that are very important for class discrimination when their expression pattern is combined with that of some other variables.[1,2] Such complementary combinations of variables can be identified only via multivariate approaches.

Another quite common misconception is to "preprocess" training data using unsupervised methods, such as clustering or principal component analysis. The idea behind this approach is to reduce dimensionality of the problem by replacing the original variables by their subset or by their combinations. However, in an unsupervised environment there is no way to determine how much of the important discriminatory information is removed by such dimensionality reduction. For example, principal component analysis identifies directions of the most variation in the data. Such directions may be very different from the direction that best separates the differentiated classes. The latter can be identified only using supervised methods.[1,3]

Nevertheless, it has to be stressed out that applying supervised learning algorithms indiscriminately – even if they are powerful methods, such as support vector machines or discriminant analysis – may lead to results that are due to random chance. In a multidimensional space of many variables, it is quite easy to perfectly or nearly perfectly separate sparse hyperareas representing different phenotypes. However, such results may just overfit the training data, and be useless as diagnostic, prognostic, or any other biomarkers.

To avoid (or at least to minimize the danger of) overfitting the training data, we should look for characteristic and repeatable expressions patterns, which – when combined into diverse multivariate biomarkers – can significantly separate the phenotypic classes. In the following sections, we will first describe how to use supervised feature selection to identify the *Informative Set of Genes*, which we define as the set containing all of the information significant for class differentiation. Then, we will present the *Modified Bagging Schema*, which utilizes the bootstrap and the ensemble approach paradigms to identify expression patterns that are most important for the class differentiation. Finally, we will show how to combine these methods to identify robust multivariate biomarkers with plausible biological interpretation.

2. The *Informative Set of Genes*

By applying a heuristic approach to multivariate feature selection, we can identify a small subset of variables, which may constitute a multivariate biomarker separating the differentiated phenotypic classes. However, a biomarker identified by a single heuristic search may be prone to overfitting and rarely provides sufficient insight into biological processes underlying class differences. To be able to find more robust biomarkers, and to have a better starting point for biological interpretation of the class differences, we identify the informative set of variables. If we focus on gene expression data, we will call such a set the *Informative Set of Genes*.

The *Informative Set of Genes* is defined as a set containing all of the information significant for the differentiation of the phenotypic classes represented in the training data.[1] To identify this set, we utilize heuristic feature selection to generate a sequence of multivariate biomarkers. After the first biomarker is identified, its variables are removed from the training data, and the second – alternative – biomarker is identified. Then, the variables of this second biomarker are also removed, and the subsequent alternative marker is identified. This process continues until no alternative biomarkers with satisfactory discriminatory power can be identified (that is, when the remaining training data no longer include any significant discriminatory information). The variables of the first biomarker and the variables selected into the identified alternative biomarkers make up the *Informative Set of Genes* (see the top part of Figure 1).

Given that the *Informative Set of Genes* contains all of the significant discriminatory information, we may assume that all gene expression patterns associated with important biological processes underlying class differences are also represented therein. To identify those patterns (and thus facilitate biological interpretation), we may utilize clustering methods, such as hierarchical clustering or self-organizing maps. Please note, however, that the unsupervised approach is applied here to the *Informative Set of Genes*, which includes only the genes whose expressions have already been determined (by supervised methods) to be associated with the class differences.

3. The *Modified Bagging Schema*

Even though the *Informative Set of Genes* contains all of the significant discriminatory information, it is possible that some of its genes were selected into some of the alternative markers due to fitting noise. To be able to find generalizable biomarkers, we need to identify the genes and the expression patterns that are most likely to be associated with biological processes responsible for class differences. To achieve this, we first build an ensemble composed of a large number of classifiers that are based on bootstrap versions of the training data, and then examine the distribution of the genes included in the *Informative Set of Genes* among the ensemble's classifiers.

Bootstrap aggregating (*bagging*) approach is commonly used to generate such ensembles of classifiers.[4] The classical version of *bagging* employs Efron's nonparametric bootstrap.[5,6] This bootstrap method makes no assumption about the investigated populations, performs selection of training samples with replacement, and produces randomized samples of the same size as the original training data. However, some learning algorithms and feature selection methods require that the observations included in a training set are independent; this independence assumption would be violated by sampling with replacement.

When heuristic feature selection is applied to each of the bootstrap training sets, its search algorithm has to compare, at each step, subsets that include the same number of genes. Obviously, this eliminates the external cross-validation approach. On the other hand, internal cross-validation that uses the same set of genes and biological samples both for training and for validation is highly unreliable when dealing with high-dimensional and sparse data. Hence, the methods of subset evaluation that are based on a metric of discriminatory power or class separation have to be used. Among them is our preferred feature selection method – stepwise hybrid search utilizing the Lawley-Hotelling T^2 criterion of class separation,[1] which requires the independence of training set observations. To achieve this independence, we modified bagging to generate randomized training sets without replacement.

We have introduced the *Modified Bagging Schema* as a method that selects bootstrap training sets by performing stratified random sampling without replacement.[1] The selection is driven by the parameter γ_{OOB} that defines a desired proportion of out-of-bag samples (training observations that are not selected into a bootstrap training set). Consequently, when the *Modified Bagging Schema* is utilized to generate hundreds or thousands of bootstrap training sets, each of such sets contains a specified proportion of biological samples randomly selected from the original training data without replacement. Consider, for example, that we generate 1,000 bootstrap training set, and use $\gamma_{OOB} = 0.8$. Then, an independent feature selection is performed for each of these bootstrap training sets, and 1,000 classifiers are built. Each of these classifiers is built on eighty percent of randomly selected training observations, and the remaining twenty percent of the training observations constitute the classifier's out-of-bag samples that can be used to estimate its classification efficiency.

4. Identification of Multivariate Biomarkers that are Robust and Biologically Interpretable

To design a biomarker discovery method capable of delivering multivariate biomarkers that are interpretable and more robust than a single biomarker selected from the entire training data, we combine the *Modified Bagging Schema* (implementing randomization and ensemble approach) with the analysis of expression patterns represented in the *Informative Set of Genes* (see Figure 1).

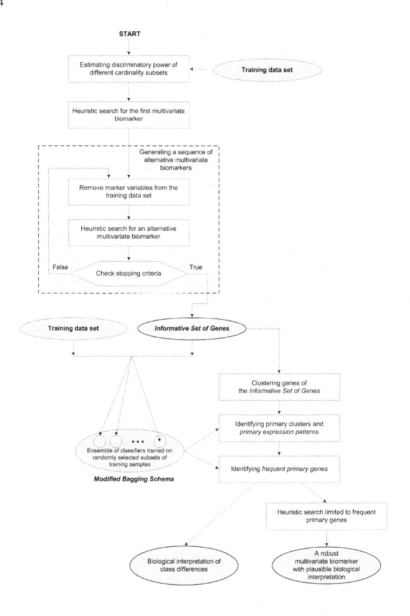

Fig. 1. Using the concepts of the *Informative Set of Genes* and *Modified Bagging Schema* to identify parsimonious multivariate biomarkers that are robust and biologically interpretable.

First, we group variables of the *Informative Set of Genes* into clusters with similar shapes of their expression patterns across all biological samples. Then, limiting our training data to the training set containing only the variables included in the *Informative Set of Genes* and utilizing the *Modified Bagging Schema*, we generate an ensemble composed of a large number of classifiers based on bootstrap training sets selected from this new training set. After that, we examine the distribution of each cluster's genes among these classifiers, and identify the primary clusters, that is, those clusters whose genes are most frequently used by the classifiers of the ensemble. For this purpose, we can consider either all of the ensemble's classifiers, or only those of them which perfectly (or nearly perfectly) classify their out-of-bag samples.

We hypothesize that the primary clusters represent the *primary expression patterns*, that is the patterns associated with the most important biological processes underlying differences among the considered phenotypic classes. Furthermore, we assume that some of the genes included in the primary clusters are more important for the class differentiation than the others. Those of them that are most frequently selected into these classifiers that perfectly or nearly perfectly classify their out-of-bag samples are deemed the *frequent primary genes*. Since they best represent the *primary expression patterns*, and since they are least likely to be selected by chance into the biomarkers included in the *Informative Set of Genes*, they constitute the best starting point for elucidation of biological processes associated with class differences. Hence, a multivariate biomarker identified by performing heuristic feature selection only on the *frequent primary genes* has the best chance of being a robust biomarker with a plausible biological interpretation.

5. A Case Study

For a case study, we selected two acute lymphoblastic leukemia (ALL) gene expression data sets (ALL3 and GSE13425)[7,8] and combined them into one training data set. Combining two data sets originated in different medical centers and representing patients from different countries (USA and Germany) should result in better representation of the targeted populations of ALL subtypes. The goal of this study was to identify a parsimonious multivariate biomarker differentiating MLL subtype of ALL from four other ALL subtypes (TEL-AML1, hyperdiploid karyotypes with more than 50 chromosomes, BCR-ABL, and the fourth group of "other" ALL samples).

Our biomarker discovery experiments were performed with the use of the *MbMD* data mining software (www.MultivariateBiomarkers.com), which includes the original implementation of the algorithms described in this paper. However, these methods may be implemented within any advanced data mining or statistical software that: (i) supports multivariate heuristic feature selection driven by a well-defined criterion of class separation (such as the Lawley-Hotelling T^2 criterion); as mentioned before, commonly used cross-validation methods are useless for feature selection performed in $p \gg n$ situations, that is, when we deal with high-dimensional data consisting of a large number of variables and relatively small number of observations, which is a typical situation for gene or protein expression data, and (ii) allows for building classifiers based on randomized training data and using such supervised learning algorithms as linear discriminant analysis (LDA) or support vector machines (SVMs).

After performing low-level preprocessing and then filtering out variables whose expression measurements were not reliable or represented experimental noise, our training data (gene expression matrix) included 6,890 variables (probe sets representing genes) and 241 observations (biological samples). Two classes were differentiated, one represented by 24 MLL biological samples, the other by 217 samples of the other four subtypes.

Following the algorithm described in Section 2, and using stepwise hybrid feature selection driven by the Lawley-Hotelling T^2 criterion of discriminatory power, we generated a sequence of alternative multivariate biomarkers. Each of them consisted of eight genes (variables), as determined by initial experiments estimating discriminatory power and class separation abilities of various cardinality markers. Discriminatory power of subsequent alternative markers was quickly decreasing, and was deemed satisfactory only for the first 22 of them. Hence, the *Informative Set of Genes* included 176 genes (see Figure 2).

To identify groups of genes with similar expression patterns, we clustered the 176 variables of the *Informative Set of Genes* using the Self Organizing Map (SOM) algorithm with a rectangular 3 by 3 topology (nine clusters) and Pearson correlation. Then, using the *Modified Bagging Schema* (described in Section 3), an ensemble of 1,000 classifiers was built. Each of these classifiers was based on a different bootstrap version of training data including only 176 genes of the *Informative Set of Genes*. By examining the distributions of the genes of the nine clusters among those of the ensemble classifiers that perfectly classified their out-of-bag samples (the perfect OOB classifiers), we identified four primary clusters representing the *primary expression patterns* of the *Informative Set of Genes*. These are the patterns that are most likely associated with the most important biological processes responsible for differences among the investigated phenotypic classes. Then, we identified twenty seven *frequent primary genes*,[1] that is, those genes of the four primary clusters that were most frequently selected into the perfect OOB classifiers (see Table 1).

---

[1] More precisely, 27 probe sets (variables) that correspond to 25 unique gene symbols.

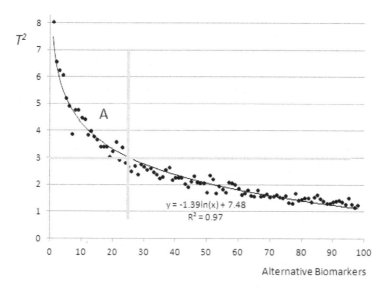

Fig. 2. The quickly decreasing discriminatory power of subsequently identified alternative biomarkers can be approximated by a logarithmic trend line. Only 22 biomarkers (represented by the points in area A) have been deemed to have satisfactory discriminatory power. Since each of them consists of eight variables, the *Informative Set of Genes* includes 176 variables.

If we now perform feature selection using only the data representing the *frequent primary genes,* we would maximize our chances for a robust multivariate biomarker. We identified two such biomarkers using two feature selection methods: T^2-driven stepwise hybrid feature selection, and recursive feature elimination with a linear support vector machine (SVM-RFE). Both biomarkers included five genes (with two genes in common).

To validate the two identified multivariate biomarkers, we selected yet another ALL gene expression data set, GSE13351.[8] Since this is an independent data set (not used for training), and since it includes patients from yet another country (Holland), the test results should be very reliable. Both our biomarkers achieved 100% MLL sensitivity. Specificity was 98.8% (85 out of 86 non-MLL patients classified correctly as non-MLL) for one of them, and 100% for the other. This indicates that our method of biomarker discovery is not only capable of finding parsimonious multivariate biomarkers that are likely to have plausible biological interpretation (since they tap into the *primary expression patterns* associated with class differences), but also highly robust ones.

Table 1. The *frequent primary genes* identified for the case study differentiating MLL subtype of ALL from four other subtypes of ALL. Since the *frequent primary genes* best represent the *primary expression patterns*, they constitute the best starting point for elucidation of biological processes associated with class differences.

Probe set	GenBank accession	Gene symbol
219463_at	NM_012261	C20orf103
209905_at	AI246769	HOXA9
203434_s_at	NM_007287	MME
201153_s_at	N31913	MBNL1
209354_at	BC002794	TNFRSF14
219686_at	NM_018401	STK32B
204069_at	NM_002398	MEIS1
201151_s_at	BF512200	MBNL1
209038_s_at	AL579035	EHD1
211066_x_at	BC006439	PCDHGC3
211126_s_at	U46006	CSRP2
208729_x_at	D83043	HLA-B
212237_at	N64780	ASXL1
208112_x_at	NM_006795	EHD1
206492_at	NM_002012	FHIT
200989_at	NM_001530	HIF1A
220448_at	NM_022055	KCNK12
200953_s_at	NM_001759	CCND2
212762_s_at	AI375916	TCF7L2
208779_x_at	L20817	DDR1
208690_s_at	BC000915	PDLIM1
219821_s_at	NM_018988	GFOD1
204304_s_at	NM_006017	PROM1
203795_s_at	NM_020993	BCL7A
218764_at	NM_024064	PRKCH
203837_at	NM_005923	MAP3K5
213150_at	BF792917	HOXA10

6. Conclusions

In the paper, we discussed common misconceptions in biomarker discovery as well as the limitations of popular feature selection approaches when applied to high-dimensional $p \gg n$ data. One needs to remember that one of the aspects of the *curse of dimensionality* is the sparsity of multidimensional representation of such data. It does mean that it is quite likely to find good separation of training samples even for data representing randomly generated noise. Although, in some situations, especially when the differentiated classes are easily separable, a single search for a multivariate biomarker may result in an efficient biomarker, it is more likely that the result of a single search overfits the training data and is of no value for classification of new samples. Nonetheless, even when a single search produces a valid biomarker, such a single biomarker rarely allows for the identification of biological processes underlying class differences.

We presented a novel method that is capable of the identification of multivariate biomarkers that are robust and interpretable. We first described two new concepts - the *Informative Set of Genes* (a set containing all of the information significant for class differentiation) and the *Modified Bagging Schema* (generating bootstrap training sets by stratified random sampling without replacement), and then showed how to combine these two concepts into a powerful biomarker discovery method. By identifying *primary expression patterns* and their *frequent primary genes*, this method leads to parsimonious multivariate biomarkers that are robust as well as having the best chance for plausible biological interpretation.

References

1. D. M. Dziuda, *Data Mining for Genomics and Proteomics: Analysis of Gene and Protein Expression Data* (Wiley, Hoboken, 2010).
2. I. Guyon and A. Elisseeff, *Journal of Machine Learning Research*, **3**, 1157 (2003).
3. D. J. Hand, H. Mannila and P. Smyth, *Principles of Data Mining* (MIT Press, Cambridge, 2001).
4. L. Breiman, *Machine Learning* **24**, 123 (1996).
5. B. Efron, *The Annals of Statistics* **7**, 1 (1979).
6. B. Efron and R. Tibshirani, *An Introduction to the Bootstrap* (Chapman & Hall, New York, 1993).
7. M.E. Ross *et al*, *Blood* **102**, 2951 (2003).
   M. L. Den Boer *et al*, *Lancet Oncol* **10**, 125 (2009).

ON SEQUENTIAL CHANGE-POINT DETECTION STRATEGIES

E. GOMBAY

Department of Mathematical and Statistical Sciences, University of Alberta,
Edmonton, Alberta, T6G 2G1, Canada
E-mail: egombay@ualberta.ca

We evaluate various sequential monitoring methods from the application point of view and consider the effect of initial parameter estimation based on a historical sample of size m. We show that the level of significance of some procedures are distorted by such estimation and propose truncated procedures where this is not the case. We also remark on the practicality of non-truncated sequential testing framework in general, and show that there is no optimal sequential test if the evaluation is considering type I error and fast stopping.

1. Introduction

In this note the problem of change point detection is considered from the practical point of view. Several strategies are considered that are easy to apply and their properties, such as Type I error probability and speed of disorder detection are compared. As applicability is our focus, some theoretically optimal procedures which have strong assumptions are omitted from the current discussion.

The term *open ended sequential tests* was defined by Robbins[24] as all those sequential tests where with a positive probability sampling never stops. In other words, tests where the stopping time can take an infinite value with probability greater than zero. Other authors use the term simply meaning that there is no maximum sample size fixed at the beginning of monitoring, hence it is just a term meaning *not truncated*, or *not curtailed*. In this discussion we shall use the term in this wider meaning, which includes tests where with probability one the stopping time is finite, but it can take any positive integer value.

Sequential monitoring has an ever growing field of applications. The first such tests, the Shewhart charts, were introduced to meet the needs of industrial quality control in the 1930's. Presently, sequential tests are also used in the medical field to monitor processes as the performance of surgeons, for example (see Gombay et al.[12] and references therein). Such tests have potential applications whenever data come in a sequence during time.

The works of Bartlett,[3] Barnard,[2] and especially Wald[27] gave a great impetus to the development of the field. The famous methods of Page,[21,22] Lorden,[17] and Moustakides[19] for sequential change point detection are related to Wald's sequential probability ratio test. These tests are open ended in the wider sense of the term, as the probability of stopping is one even under the null hypothesis of no change, which is saying, of course, that the probability for Type I error is one. On the other hand, they have the celebrated optimality of fast stopping. More precisely, these tests stop fastest conditionally, given that the change point has been passed. Mathematically, let $X_1, X_2, \ldots$ be independent observations, such that $X_1, \ldots, X_{\tau-1}$ have d.f. F_0, and $X_\tau, X_{\tau+1}, \ldots$ have d.f. F_1, $(F_0 \neq F_1)$, with a finite change point τ, assumed to be unknown, and let $E_\tau(.)$ denote expectation under the distribution when change is at epoch τ. The optimal stopping rule N of Lorden[17] minimizes

$$\sup_{\tau \geq 1} ess\, sup\, E_\tau\{(N - \tau + 1)^+ | X_1, \ldots, X_{\tau-1}\} \tag{1}$$

among all stopping rules which have expected value of stopping time under the no-change hypothesis H_0, $(\tau = \infty)$, greater than or equal than some design parameter γ, $(\gamma > 0)$, that is $E_\infty(N) \geq \gamma$.

These tests have come under some scrutiny lately. In Mei [18] it has been shown that the control limits of the monitoring algorithm may work quite differently than expected under many scenarios, resulting in performances that are not well understood. Furthermore, although this optimality property is theoretically very appealing, and it gives best possible results on the speed of stopping after change, from the practical point of view it is not always helpful as pointed out in Gombay et al.[12] and Gombay and Serban.[13] The problem is that the optimality is conditional on having passed the true change-point, but these tests will often stop before change due to the fact, that the distribution of the stopping time under the hypothesis of no change is strongly skewed. So these tests are really good and reliable only for very early changes in the sequence of observation, but are of questionable reliability, if changes are occurring later. They can, however, serve as benchmarks to evaluate other change detection tests.

We do not consider Bayesian change point detection algorithms in this note, although in those both type I error rates and fast stopping are used as measures of performance, and they can have optimally short detection delay. However, there are distributional assumptions used in them that we do not have, and these can result in unjustified comparisons. As Lai and Xing[15] note, there is "considerable cost due to ignorance" of parameters. For a good review of Bayesian procedures see Tartakovsky and Moustakides.[25]

Another type of open ended monitoring method has emerged in the paper Berkes et al.[4] This idea was used by several authors (see Horváth et al.,[14] Aue et al.,[1] Lee et al.,[16] and references therein) for various models. In this note we discuss and analyze algorithms, and argue that the open ended tests are not effective and powerful for sequential change point detection unless the change is early in the sequence. This is, of course, a counter indication for testing for a long indefinite period.

Truncated sequential tests for monitoring can be defined, and they will be compared to the above tests, and we conclude, that they are better in some respects than the open ended ones. The choice of the truncation parameter is also discussed.

2. An Open Ended Strategy

As in Chu et al.,[5] for the simplicity of exposition, and to emphasize the essence of our message, we assume $X_1, X_2, \ldots$ to be independent observations. However, due to invariance principles the conclusions will be valid for many other models as well, as long as approximating the stochastic process by a Brownian motion is possible. Such models include linear regression as in Chu et al.,[5] GARCH time series as in Berkes et al.,[4] multiparameter exponential family of distributions as in Gombay,[9] stationary time series as in Gombay and Serban,[13] for example. We concentrate on the change in mean, and we assume that the variance of the observations is one for all observations. Change detection methods test the hypothesis

$$H_0 : E(X_i) = \mu_1, \ i = 1, 2, \ldots$$

against the alternative, that at some unknown time τ the mean changes, that is, against the hypothesis

$$H_A : E(X_i) = \mu_1, \ i = 1, 2, \ldots, \tau - 1, \ E(X_i) = \mu_2, \ i = \tau, \tau + 1, \ldots,$$

where $\mu_1 \neq \mu_2$. In Berkes et al.[4] the following type of tests were introduced. In that setting $\tau > m$ is assumed and $X_1, \ldots, X_m$ denotes the historical sample that is gathered before the start of testing. The decision rule is, stop and declare that change happened as soon as the following event is observed:

$$
\sup_{1 \le k < \infty} m^{-\frac{1}{2}} \left| \sum_{m < i \le m+k} X_i - \frac{k}{m} \sum_{1 \le i \le m} X_i \right| \Big/ \left(1 + \frac{k}{m} \right)
$$

$$
= \sup_{1 \le k < \infty} m^{-\frac{1}{2}} \left| \sum_{m < i \le m+k} (X_i - \bar{X}_m) \right| \Big/ \left(1 + \frac{k}{m} \right) \ge b, \qquad (2)
$$

for some constant $b = b(\alpha)$, chosen to make the type I error equal to α. We note, that instead of the constant b one can take an appropriately chosen function $b = b(\frac{k}{m})$, but we shall not pursue this modification here.

The motivation is clear. The historical sample $X_1, \ldots, X_m$ specifies the acceptable state of nature, and change from it has to be detected. In the retrospective change detection methodology (2) appears with standardization, the ranges of supremum and summations being different in the stopping rule:

$$
\sup_{1 \le k \le m} m^{-\frac{1}{2}} \left| \sum_{1 < i \le k} X_i - \frac{k}{m} \sum_{1 \le k \le m} X_i \right| = \sup_{1 \le k \le m} m^{-\frac{1}{2}} \left| \sum_{1 \le k \le m} (X_i - \bar{X}_m) \right| \ge c.
$$
$$
(3)
$$

Note, that in both cases standardization involves $m^{-1/2}$, but the role of m is different in the two cases. It is the historical sample size in the sequential (online) test; and it is the sample size in the retrospective, or offline change detection. In both cases, the critical levels b and c, respectively, are calculated from asymptotic distributions as $m \to \infty$. Consistency of such tests in the retrospective change detection makes a clear sense, but in the sequential situation it is somewhat questionable, because m is fixed and the sample size is random once testing has been started. Furthermore, we shall see that continuing the test longer may not lead to the rejection of H_0 with probability one even under the alternative hypothesis H_A. Hence, consistency here has a different meaning than usually understood. We note, that Chu et al.[5] also used asymptotic distribution with the historical sample size m converging to infinity. We shall come back to the effect of this in our discussion on the performance of these tests when initial parameters have to be estimated based on the available data set of size m.

For the test in (2) the critical value b is derived using the invariance principle for independent and identically distributed random variables as follows. If μ denotes the true mean of the observations, then for a T fixed, as $m \to \infty$

$$m^{-\frac{1}{2}} \sum_{1 \leq i \leq mt} (X_i - \mu) \to^D W(t), \ 0 \leq t \leq T,$$

where $W(t)$ denotes a standard Brownian motion. Hence for $0 \leq t \leq T$

$$m^{-\frac{1}{2}} \left| \sum_{m \leq i \leq m(1+t)} (X_i - \mu) - t \sum_{1 \leq i \leq m} (X_i - \mu) \right| \to^D |W(1+t) - (1+t)W(1)|,$$

as $m \to \infty$. By the law of iterated logarithm for the Brownian motion, as $T \to \infty$

$$\sup_{T \leq t < \infty} \frac{W(1+t)}{1+t} \to^{a.s.} 0.$$

Assuming $E|X_1 - \mu|^p < \infty$ for some $p > 2$, the partial sum process $\sum_{i=1}^n (X_i - \mu)$ can be approximated by a Brownian motion with *almost sure* error rate $O(n^{-1/p})$, and using this we arrive at the following convergence result:

$$\sup_{1 \leq k < \infty} \frac{|\sum_{m < i \leq m+k}(X_i - \mu) - \frac{k}{m}\sum_{1 \leq i \leq m}(X_i - \mu)|}{m^{1/2}(1 + \frac{k}{m})}$$

$$\to^D \sup_{0 < t < \infty} \frac{|W(1+t) - (1+t)W(1)|}{1+t} =^D \sup_{0 < s < 1} |W(s)|, \ m \to \infty. \quad (4)$$

The last equality in distribution can be seen by calculating covariance structures.

To see the behavior of the statistic in (2) under the change alternative H_A it is useful to write

$$\sqrt{m}(\hat{\mu}_{m+k} - \hat{\mu}_m) = \sqrt{m} \left\{ \frac{\sum_{1 \leq i \leq m+k} X_i}{m+k} - \frac{\sum_{1 \leq i \leq m} X_i}{m} \right\}$$

$$= \sqrt{m} \left\{ \frac{\sum_{1 \leq i < \tau} X_i + \sum_{\tau \leq i \leq m+k} X_i}{m+k} - \frac{\sum_{1 \leq i \leq m} X_i}{m} \right\}$$

$$= \sqrt{m} \left\{ \frac{\sum_{1 \leq i < \tau}(X_i - \mu_1) + \sum_{\tau \leq i \leq m+k}(X_i - \mu_2)}{m+k} - \frac{\sum_{1 \leq i \leq m}(X_i - \mu_1)}{m} \right\}$$

$$+ \sqrt{m} \left\{ \left(\frac{\tau - 1}{m+k} - 1 \right) \mu_1 + \frac{m+k-(\tau-1)}{m+k} \mu_2 \right\} = A^{(1)} + A^{(2)}. \quad (5)$$

$A^{(1)} = A^{(1)}(m, k)$ has the same asymptotic distribution as (2) under H_0, and $A^{(2)} = A^{(2)}(\tau, m, k)$ is the drift. In a given test m, τ are fixed, as m is the historical sample size, and τ is the unknown non-random change point, so as k, the sample size, increases the drift converges to $m^{1/2}(\mu_2 - \mu_1)$, showing that the test cannot be consistent in the usual sense where the power converges to one as sample size k converges to infinity. Furthermore, as the increase in the drift with an additional sample at stage k is $O(k^{-2})$ only, and $A^{(1)}$ has the distribution of $\sup_{0<s<1} W(s)$ approximately, continuing testing for a longer period in this open ended scenario has little effect on the power.

Note, that if the monitoring with (2) were restricted to the range $\{k : 1 \leq k \leq jm\}$ for some fixed j, then due to the time transformation $s = t/(1+t)$, to obtain the critical value b the range can be restricted to

$$\sup_{0<s<\frac{j}{j+1}} |W(s)| = \sqrt{\frac{j}{j+1}} \sup_{0<u<1} |W(u)|,$$

which gives a sharper critical value, hence it will increase the power. An additional advantage of using $n = jm$ for the truncation point is that it validates the asymptotical consideration that prove the consistency of the tests.

In Chu et al.[5] and Gombay[9] open ended tests of change based on Robbins' (1970)[24] ideas were discussed. For our simple model one such stopping rule is

$$\sup_{k\geq 1} \sum_{1\leq j\leq k} (X_i - \mu_1) \geq [(k+1)(a^2 + \log(1+k))]^{1/2}, \tag{6}$$

with $a = (-2\log\alpha)^{1/2}$ for α level of significance, and μ_1 the in-control mean value. It is easy to see that this test is consistent in the sense that for any fixed size of change, at any time τ, the power will be one if monitoring continues indefinitely. Note that this is a sequential test by design, but it can also serve for change detection. In Gombay[9] a small simulation study where variance was estimated, demonstrated that the power is one for any location of change if monitoring continues indefinitely. It is easy to see, that if μ_1 is not known, but it is estimated by $\bar{x}_m$, then the type one error will be one for any m as the Central Limit Theorem gives $O_P(m^{-1/2})$ fixed size of error. The limit theorems with $m \to \infty$ mask this problem, but in practical situations m is fixed in each application once testing has started.

3. Truncated Procedures

As monitoring indefinitely may be not practical in many applications, and estimation may distort the size of the open tests, sequential truncated tests offer a reasonable alternative. One such strategy is Test 3 of Gombay,[9] where a maximal sample size, or truncation point n has to be chosen at the start of testing. The stopping rule is as follows.

If for some k, $1 < k \leq n$,

$$STAT(k) = n^{-1/2} \max_{1 \leq j \leq k} \sum_{j \leq i \leq k} (X_i - \mu_1) \geq F_\alpha, \tag{7}$$

then stop and reject the null hypothesis, otherwise, if $k < n$, then take another sample. If $STAT(k) < F_\alpha$ for all $k \leq n$, then no evidence against the null hypothesis has been found.

F_α is obtained from the distribution of $\sup_{0 < u < 1} |W(u)|$. In this test statistic the partial sum is of the same structure as Page's, with different standardizing due to truncation, so (7) is the truncated version of Page's test.

If change is at a point $\tau = \gamma n$, $0 < \gamma < 1$, then the test is consistent in the usual sense as $n \to \infty$, as the drift is

$$n^{-1/2} \max_{\tau \leq i \leq k \leq n} \sum_{\tau \leq i \leq k} (\mu_2 - \mu_1).$$

The choice of truncation point n can be made using arguments similar to the choice of sample size n in non-sequential setting. By the nature of the problem, we need, of course, to fix more parameters. Let α be the level of significance, $1 - \beta$ the required power at an increase of the mean of size $d = \mu_2 - \mu_1$. If F_α denotes the upper α-quantile of the distribution of $\sup_{0 < u < 1} |W(u)|$, and we wish to have power $1 - \beta$ after $k = \delta n$ $(0 < \delta < 1)$ observations have been taken after change, then routine calculations give the choice of a truncation point

$$n = \frac{(F_\alpha - F_{1-\beta})^2}{\delta d^2}.$$

The achieved power will be more or less than $1 - \beta$, depending on $\delta + \gamma \leq 1$ or $\delta + \gamma > 1$, respectively. When the variance is not known, d/s has to be used instead of d in the formula, where s^2 is the estimated variance based on historical data. These simple calculations show that if $\delta + \gamma \leq 1$, then with probability $1 - \beta$, the delay time is no more than $O_P(\delta n)$. For example, one may wish to have power of 80 % after of $(0.1)n$ observations after change, and the size of departure from $\mu_1 = 0$ that is of special concern is $\mu_2 = 0.25$. Then assuming variance one and $\alpha = 0.05$ $n = 323$ is sufficiently large for our purposes. (We used $F_{0.05} = 2.24$ and $F_{0.80} = 0.82$.)

Similar to (6), sequential tests Test 1 and 2 of Gombay[9] can also detect changes. As in (2) and (6) the cumulative sums are taken from the beginning of the monitoring. Test 1 has the following stopping rule.

If for some k, $1 < k \le n$,

$$STAT(k) = \frac{1}{\sqrt{k}} \left| \sum_{1 \le i \le k} (X_i - \mu_1) \right| \ge C_1(\alpha, n), \qquad (8)$$

then stop and reject the null hypothesis, otherwise, if $k < n$, then take another sample. If $STAT(k) < C_1(\alpha, n)$ for all $k \le n$, then no evidence against the null hypothesis has been found.

Critical value $C_1(\alpha, n)$ can be approximated by the results of Vostrikova[26] for each n. Her approximations give less conservative critical values for finite sample sizes, than the asymptotic double exponential distribution, as convergence for extremal values is known to be very slow. Her approximation for dimension one is

$$P\{ \sup_{1 < k \le n} k^{-\frac{1}{2}} | \sum_{1 \le i \le k} (X_i - \mu_1)| > y\} \cong \frac{y \exp(-\frac{1}{2}y^2)}{\sqrt{2\pi}} \{T(1 - \frac{1}{y^2}) + \frac{4}{y^2} + O(\frac{1}{y^4})\},$$

so $C_1(\alpha, n)$ can be obtained from this equation with $T = \log n$. Note, that the statistic in (8) is the sequential version of the usual fixed sample size statistic $S_n/\sqrt{n}$, $S_n = \sum_{1 \le i \le n} X_i$, and of Pocock's[23] group- sequential testing scheme.

Test 2 was defined by stopping rule:

If for some k, $1 < k \le n$,

$$STAT(k) = \frac{1}{\sqrt{n}} | \sum_{1 \le i \le k} (X_i - \mu_1)| \ge F_\alpha, \qquad (9)$$

then stop and reject the null hypothesis, otherwise, if $k < n$, then take another sample. If $STAT(k) < F_\alpha$ for all $k \le n$, then no evidence against the null hypothesis has been found.

Note, that statistic process (9) defines the continuous version of the O'Brien-Fleming[20] group-sequential tests.

4. Theoretical Comparisons

The asymptotic distribution to which Vostrikova[26] is an approximation, is the Darling and Erdős[6] result for independent, identically distributed random variables $\{X_i\}$ with mean zero and variance one:

$$\lim_{n\to\infty} P\{\max_{1\le k\le n} k^{-1/2}a(n)(|\sum_{1\le i\le k} X_i|) - b(n) \le y\} = \exp(-2e^{-y}), \quad (10)$$

for $-\infty < y < \infty$, where $a(n) = (\log\log n)^{1/2}$, $b(n) = 2\log\log n + 1/2\log\log\log n - 1/2\log\pi$. Following Gombay[8] in the Appendix we show that the statistic of (8) takes its maximum early in the sequence of observations, and it is not sensitive to departures from the true mean that are of order $O(n^{-1/2})$.

This means that test (8) is not sensitive to small deviation from the null value of the mean μ_1, but stops earlier than (9). In reality, however, μ_1 is often not known, but it is estimated from a historical sample of size m as in (2). If we assume that m is proportional to the maximum sample size n, then by the CLT the error of approximation $\mu_1 - \bar{x}_m = O(n^{-1/2})$. Hence the level of (9) and (7) are distorted, but the performance of (6) and (8) are not affected by this approximation according to large sample approximations.

So (7) and (9) can be used only if the initial true parameter, in our example μ_1, is known. Such situations occur, for example, in industrial monitoring, when the null hypothesis specifies the true target in manufacturing. In such applications they can outperform (8) in certain scenarios as shown in Gombay.[9] In case of estimated initial parameter value (7) and (9) do not give reliable procedures. Naturally, the larger m is relative to n, the less distortion this approximation makes in the level of the tests. Simulation studies confirm these theoretical finding.

5. Empirical Comparisons

The purpose of this section is to demonstrate the properties of some tests when initial parameter estimation is necessary. We present detailed results only of those strategies that work under such circumstances, namely tests based on (2) open, (2) truncated, and (8).

Simulation studies, not presented here, confirm that all other tests have distorted empirical levels, giving much higher type one error than the nominal α, so we do not include those in our tables below. We note, that Chu et al.[5] proposed an open ended sequential testing strategy for monitoring structural change in the parameters of linear regression. A historical period of length m was used to estimate the in-control parameter value. In our simple example of testing for change in the mean this strategy corresponds to monitoring $\sum_{i=1}^{k}(X_i - \bar{X}_m)$. Note, that the error relative to test (6) with known initial parameter value at stage k is $\sum_{i=1}^{k}(\mu_1 - \bar{X}_m) = kO_P(m^{-1/2})$. Limit theorems derived as $m \to \infty$ show the consistency of tests in Chu et al.,[5] but in practice m the historical sample size is always fixed once testing has started, and as the current sample size k converges to infinity the type I error will be one for any fixed m. This property of (6) was confirmed in simulations with in control parameter $\mu_1 = 0.0$ estimated based on a sample of size m. Under the no change situation it can be as high as $\hat{\alpha} = 0.7290$, which leads to the distortion of the empirical power as well. On the other hand, in Gombay [9] this test was shown to perform well if the mean is known and the variance is estimated.

In our simulation studies we consider testing for H_0 when neither the mean of the variance is given. As in Gombay et al.[11] one can show that the level of test is not distorted if the standard deviation is consistently estimated. The proof of the fact that, asymptotically, the truncated test of (8) is insensitive to the estimation of the mean can be found in the Appendix.

In Table 1 we demonstrate that although both the open and truncated versions of tests based on (2) have a good control over the empirical level, tests based on (8) need a good approximate value for the mean. In the various simulations performed we find that the initial sample size m has to be at least as large as the truncation point in the current scenario of testing for change in the mean. This is because the results in the Appendix are using large sample approximations. Different models can have various minimum requirements of the size of m relative to n. In Gombay et al.[12] the initial covariate parameters of the logistic model were estimated on two years' data and the monitoring went on for four years. Simulation for this model showed excellent empirical level for such ratio of m and n. Hence, in applications Monte Carlo experiments should be performed for tests based on strategy like (8) to have a good idea what size of m is needed. Note, that there are cases, when testing does not require such estimation, even in the absence of exact initial parameter values. Such test at are two sample tests as in Gombay et al.;[11] tests based on anti-symmetric U-statistics as in Gombay;[7] tests using the logrank statistics as in Gombay.[10]

Tables 2 and 3 show that there is no optimal test, as the tests outperform each other in different situations. Test (8) stops the fastest if the change is early in the sequence and the size of the change is not too small. The

Table 1.   Empirical levels for various truncation points (n) and historical sample sizes (m).

Sample size	(2) open[a]	(2) truncated	(8) truncated
$n = 2000,\ m = 500$	0.059	0.06	0.260
$n = 2000,\ m = 1000$	0.043	0.052	0.129
$n = 2000,\ m = 2000$	0.032	0.053	0.077
$n = 2000,\ m = 3000$	0.013	0.057	0.066
$n = 2000,\ m = 4000$	0.012	0.049	0.047

Note: [a] Open ended test continued to $8,000$ observations.

Table 2.   Empirical power for various change-points τ and alternative means μ_2. Truncation point is $n = 500$ and historical sample size $m = 2000$. Initial true mean $\mu_1 = 0.0$.

	(2) open[a]	(2) truncated	(8) truncated
$\tau = \infty,\ \mu_2 = 0.0$	0.0320	0.0575	0.0435
$\tau = 1,\quad \mu_2 = 0.1$	0.952	0.480	0.323
$\tau = 1,\quad \mu_2 = 0.2$	1.000	0.962	0.906
$\tau = 1,\quad \mu_2 = 0.3$	1.000	1.000	0.999
$\tau = 1,\quad \mu_2 = 0.4$	1.000	1.000	1.000
$\tau = 1,\quad \mu_2 = 0.5$	1.000	1.000	1.000
$\tau = 100,\ \mu_2 = 0.1$	0.943	0.3265	0.1740
$\tau = 100,\ \mu_2 = 0.2$	1.000	0.8475	0.6915
$\tau = 100,\ \mu_2 = 0.3$	1.000	0.994	0.974
$\tau = 100,\ \mu_2 = 0.4$	1.000	1.000	1.000
$\tau = 100,\ \mu_2 = 0.5$	1.000	1.000	1.000
$\tau = 200,\ \mu_2 = 0.1$	0.936	0.187	0.094
$\tau = 200,\ \mu_2 = 0.2$	1.000	0.583	0.391
$\tau = 200,\ \mu_2 = 0.3$	1.000	0.912	0.802
$\tau = 200,\ \mu_2 = 0.4$	1.000	0.994	0.975
$\tau = 200,\ \mu_2 = 0.5$	1.000	1.000	0.9995

Note: [a] Open ended test continued to $8,000$ observations.

open ended test has the greatest power for small changes at the expense of a large average stopping time. Its truncated version performs well in most situations.

In the simulations presented the nominal level of significance was $\alpha = 0.05$, and observations had normal distribution. The empirical power values were calculated based on 2000 repetitions for each scenario.

We note, that these tests are based on asymptotics, and due to invari-

Table 3. Empirical average stopping times (with standard deviations) for various change-points τ and alternative means μ_2. Truncation point is $n = 500$ and historical sample size $m = 2000$. Initial true mean $\mu_1 = 0.0$.

	(2) open[a]	(2) truncated	(8) truncated
$\tau = \infty$, $\mu_2 = 0.0$	7876(755)	493(36)	487(70)
$\tau = 1$, $\mu_2 = 0.1$	2455(1829)	426(99)	425(135)
$\tau = 1$, $\mu_2 = 0.2$	696(231)	263(98)	234(146)
$\tau = 1$, $\mu_2 = 0.3$	412(96)	168(52)	111(76)
$\tau = 1$, $\mu_2 = 0.4$	293(55)	123(32)	62(40)
$\tau = 1$, $\mu_2 = 0.5$	226(37)	96(22)	41(25)
$\tau = 100$, $\mu_2 = 0.1$	2678(1885)	462(71)	470(86)
$\tau = 100$, $\mu_2 = 0.2$	429(256)	362(94)	381(117)
$\tau = 100$, $\mu_2 = 0.3$	531(112)	276(67)	277(92)
$\tau = 100$, $\mu_2 = 0.4$	406(66)	228(44)	218(60)
$\tau = 100$, $\mu_2 = 0.5$	337(46)	202(33)	187(43)
$\tau = 200$, $\mu_2 = 0.1$	2885(1927)	482(48)	483(73)
$\tau = 200$, $\mu_2 = 0.2$	967(2285)	439(71)	454(86)
$\tau = 200$, $\mu_2 = 0.3$	652(125)	380(70)	398(91)
$\tau = 200$, $\mu_2 = 0.4$	521(78)	335(55)	335(76)
$\tau = 200$, $\mu_2 = 0.5$	449(55)	306(41)	309(60)

Note: [a] Open ended test continued to 8,000 observations.

ance principles our findings are relevant for other distributions and other models whenever stochastic processes can be approximated by Browninan motions with small enough estimation error. As the error will be of different magnitude for the various scenarios, the proportion of m and n when the truncated tests work will vary. Hence, simulation studies under the null hypothesis of no change are very useful tools to confirm that the chosen values of m and n are appropriate.

6. Conclusions

In this note sequential change detection methods were considered from the application point of view.

The surveyed open ended monitoring procedures can fail for various reasons:

(1) The Type I error is 1;

(2) The power is not increasing with continued monitoring;

(3) Parameter estimation based on historical sample size results in in-

creased Type I error.

Our results show that the open ended change detection strategies (1) and (2) are not very good for change point detection, unless the change is early in the sequence, which, of course, is a counter indication for monitoring for an extended period of time.

Truncated tests offer a reasonable alternative, especially as all test are stopped at some finite time eventually. Monte Carlo studies show that the performance of the strategies depend on the size and on the place of change, so one cannot talk about one test being optimal uniformly for all scenarios. The two criteria used in sequential test evaluations are the error probability and the size of delay. These work contrary to each other, so they have to be balanced. It is shown, that the procedures with optimally low delay time have undesirable properties, which make their application unpractical. However, they provide the best possible conditional delay, which can be used as a benchmark. We note, that increase in stopping time is beneficial if estimation is performed after the test has been concluded. We summarize our findings in the following table.

Table 4. Summary of the properties of the various sequential tests, when the initial parameter has to be estimated.

Test	(6)	(2)	(7)	(8)	(9)
truncation	no	no	yes	yes	yes
empirical level	1.0	α	1.0	α	1.0
power	1.0	< 1.0	1.0	1.0	1.0
			$(n \to \infty)$	$(n \to \infty)$	$(n \to \infty)$

Appendix: Asymptotic Properties

First we show, that when there is no disturbance stochastic process (8) takes its maximum early in the sequence $k = 1, 2, \ldots, n$. By (10) with $c(n) = n/\log n$

$$\lim_{n \to \infty} P\{ \max_{1 \leq k \leq c(n)} k^{-1/2} a(n) | \sum_{1 \leq i \leq k} X_i - \mu_1 | - b(n) \leq y \} = \exp(-2e^{-y}),$$

$$-\infty < y < \infty.$$

By the stationarity of $\sum_{1 \leq i \leq k} (X_i - \mu_1)/k^{1/2}$ in case $\{X_i\}$ are normal random variables, and approximate stationarity in the general case (due to invariance principles), we have

$$\lim_{n \to \infty} P\{ \max_{c(n) \leq k \leq n} k^{-1/2} a(n) | \sum_{1 \leq i \leq k} X_i - \mu_1 | - b(n) \leq y \} =^D$$

$$\lim_{n\to\infty} P\{\max_{1\le k\le \log n} k^{-1/2} a(n)| \sum_{1\le i\le k} X_i - \mu_1| - b(n) \le y\} \to^P -\infty.$$

So we have

$$\max_{1\le k\le n} k^{-1/2}| \sum_{1\le i\le k} X_i - \mu_1| =$$

$$\max\{\max_{1\le k\le c(n)} k^{-1/2}| \sum_{1\le i\le k} X_i - \mu_1|, \max_{c(n)<k\le n} k^{-1/2}| \sum_{1\le i\le k} X_i - \mu_1|\} =^D$$

$$\max_{1\le k\le c(n)} k^{-1/2}| \sum_{1\le i\le k} X_i - \mu_1|.$$

It is easy to see, that in (9), where standardization is by $n^{-1/2}$, the maximum takes its value later than in interval $(1, c(n))$, as $\max_{1\le k\le c(n)} n^{-1/2}| \sum_{1\le i\le k} X_i - \mu_1| = o_p(1)$.

Next we show, that process (8) is not sensitive to small deviations from the null distribution, that is, if we add and error of size $O(n^{-1/2})$ to each observation in (8) making it $k^{-1/2} \sum_{1\le i\le k}(X_i - \mu_1 + dn^{-1/2})$, d a constant. The large sample approximation does not change from the no disturbance case because the error in the interval $[1 \le k \le \log c(n)]$ is at most a constant times $(\log\log n)^{1/2} c(n)^{1/2} n^{-1/2} = o(1)$. In the range $[c(n) \le k \le \log n]$ the total error is no more than $O(1)$, and we have

$$\lim_{n\to\infty} P\{\max_{1\le k\le \log\log n} k^{-1/2} a(n)(| \sum_{1\le i\le k}(X_i - \mu_1)| + O(1)) - b(n) \le y\} \to^P -\infty,$$

due to the fact that the convergence to $-\infty$ is coming from the leading term of $b(n)$ which converges faster to infinity than $a(n)$. Again, it is easy to see, that an $O(1)$ size error distorts the null distribution of (9), which is known to be sensitive to small changes, that is, to the so-called contiguous alternatives, when the error added to each observation is $O(n^{-1/2})$.

Acknowledgements

This research was supported in part by a NSERC Canada Discovery Grant.

References

1. A. Aue, L. Horváth, and M. L. Reimherr, *J. of Econometrics* **149**, 174 (2009).
2. G. A. Barnard, *J. Amer. Statist. Ass.* **42**, 658 (1947).
3. M. S. Bartlett, *Proc. Camb. Phil. Soc.* **42**, 239 (1946).

4. I. Berkes, E. Gombay, L. Horváth, and P. Kokoszka, *Econometric Theory* **20**, 1140 (2004).

5. C.-S. Chu, M. Stinchcombe and H. White, *Econometrica* **64**, 1045 (1996).

6. D. A. Darling and P. Erdős, *Duke Mathematical Journal* **23**, 143 (1956).

7. E. Gombay, *Metrika* **54**, 133 (2000).

8. E. Gombay, *Theory of Stochastic Processes (Kiev)* **8**, 107–118 (2002).

9. E. Gombay, *Sequential Analysis* **22**, 203 (2003).

10. E. Gombay. *Sequential Analysis* **27**, 97 (2008).

11. E. Gombay, G. Heo and A. Hussein, *Sequential Analysis* **26**, 71 (2007).

12. E. Gombay, A. Hussein and S. Steiner, *Statistics in Medicine* **30**, 2815 (2011).

13. E. Gombay and D. Serban, *Journal of Multivariate Analysis* **100**, 715 (2009).

14. L. Horváth, P. Kokoszka and J. Steinebach, *Journal of Statistical Planning and Inference* **126**, 225 (2004).

15. T. L. Lai and H. Xing, *Sequential Analysis* **29**, 162 (2010).

16. S. Lee, J. Ha, O. Na and S. Na, *Scandinavian J. of Statist.* **30**, 781 (2009).

17. G. Lorden, *Ann. Math. Statist.* **42**, 1897 (1971).

18. Y. Mei, *Sequential Analysis* **27**, 354 (2008).

19. G. V. Moustakides, *Ann. Statist.* **14**, 1379 (1986).

20. P. C. O'Brien and T. R. Fleming, *Biometrics* **35**, 549 (1979).

21. E. S. Page, *Biometrika* **41**, 100 (1954).

22. E. S. Page, *Biometrika* **42**, 523 (1955).

23. S. J. Pocock, *Biometrika* **64**, 191 (1977).

24. M. Robbins, *Ann. Math. Statist.* **41**, 1397 (1970).

25. A. G. Tartakovsky and G. V. Moustakides, *Sequential Analysis* **29**, 125 (2010).

26. L. J. Vostrikova, *Theory of Probability and Its Applications* **26**, 356 (1981).

27. A. Wald, *Sequential Analysis* (Wiley, New York, 1947).

MULTI-ITEM TWO-STAGE EOQ MODEL INCORPORATING QUANTITY AND FREIGHT DISCOUNTS

DEEPALI GUPTA

Department of Mathematics, Jaypee Institute of Information Technology,
Noida, 201307, U.P., India
E-mail: deepali.8@gmail.com

KANIKA GANDHI; SANDHYA MAKKAR; P. C. JHA

Department of Operational Research,
Faculty of Mathematical Sciences, University of Delhi,
Delhi, 110007, India

Supply chain management is an integrating function with primary responsibility for linking major business functions and business processes within and across companies into a cohesive and high-performing business models. Managing supply chain is a difficult task because of complex interrelations and integration of various entities that exist in it. One of the entities is transportation of items from source to destination. This process becomes tedious, when items are moving with one or more stoppage as on stoppage point inventory carrying cost would also be incurred. A single source, single destination and multi-item EOQ model has been discussed in the literature for single stage as well as two or more stages. Different discount policies are offered to procure and transport goods from one stage to the other. In the present study a mathematical model is developed to minimize the sum of inventory, procurement and transportation costs when several items are shipped from a single source to single destination through an intermediate stoppage, in which it is assumed that inventory carrying charge at the intermediate stoppage, is very high after a pre-specified time. A method of solution is proposed and verified numerically.

1. Introduction

The rapid evolution of retail industry has observed that mastering supply chain dynamics is critical for the growth of modern trade. In recent competitive pressures, a variety of research papers have contributed significantly in the study of two stage supply chains, integrated discounts and transportation schemes. Kaminsky and Levi[2] developed two stage production-transportation model featured capacitated production in two stages, and a fixed cost (or concave cost) for transporting the product between the stages. They show that their model reduces to a related model, with one capacitated production stage with linear production cost, and transportation between two inventory locations with non-linear transportation cost. However, another model of a two-stage push-pull production distribution supply chain is considered by Ahn and Kaminsky[1]. In their model, orders arrive at the final stage according to a Poisson process. Two separate operations, which take place at different places with exponential service times, were located to convert the raw materials into finished goods. When the first operation is completed the intermediate inventory is held at the first stage and then transported to the second stage where the items are produced to order. A two stage supply chain network of an automobile company is integrated by Subramanya and Sharma[4] which measured the performance parameters and established the priority decision and queuing rules for improving the utilization of resources. The study restricted to measuring operational processes in a two stage supply chain between the supplier - manufacturer - distributors. Further the dynamics of a two stage supply chain consisting of one retailer and one distributor with order-up-to control policy is examined by Wang et al.[5]. A two stage inventory models over an infinite planning horizon with constant demand rate and two modes of transportation is discussed by Rieksts and Ventura[3]. These transportation options include truckloads and a less than truck- load carrier. An optimal algorithm is derived for a one-warehouse one-retailer system. A power-of-two heuristic algorithm is also proposed for a one-warehouse multi-retailer system.

In recent years the electronic industry is growing at a brisk pace globally, and the demand in the Indian market is emerging, therefore investments are flowing into augment manufacturing capacity. The growth has attracted global players to India like Apple, LG, Samsung and many more have made large investments to access the Indian market. The product range in this sector comprises of official products, home appliances, communications, as well as entertainment. These can be classified into personal computers & Laptops, telephones, calculators, playback, Air-conditioners(AC), digital video disk (DVD), video compact disc (VCD), video home systems (VHS), home theatre, music players, LCDs, cameras, camcorders, portable audio etc. Selling expensive products to Indian consumer is a challenge because they are highly price sensitive and keep hopping from one place to another for good deals. At the same time companies also like to raise their performance with abridged cost. Therefore multi brand retail stores like eZone, Tata-Croma, Next, Reliance etc, came into existence who keeps their own brand as well as other branded products. These multi brand stores have distinct advantage over the smaller stand alone stores, and have the advantage of bargaining good discounts from the suppliers, as these companies have wide network and are able to capture big market share. Electronic industry is technology driven with respect to time, so companies need to constantly improvise and innovate their supply chain.

In this article, we have considered a multi brand retail store ABC in India who is dealing in electronic products like AC, LCD, Digital Camera, Laptop and I-pod that are ordered from XYZ supplier and the inventory is kept at warehouse near to the retail store. The supplier is providing discounts on order beyond a fixed quantity level. The mode of transporting the goods from supplier to retail store takes place in two stages. In the first stage, goods move using cargo; here retail store avails discounts on bulk transportation. The unloading point of goods is the intermediate stoppage. At the stoppage, unloading of goods and their further processing takes a specified time for which the holding cost is free. Since cargo point requires space to unload goods of number of cargos, the holding cost at stoppage increases with very high rates after a preset time. Keeping inventory for long time may not be beneficial for the retailer but sometimes the inventory has to be kept because of some undue cause like transportation facility is not available to move goods from cargo point to destination. Movement of goods from intermediate stoppage to retail store is the second stage of the model which is completed through modes of transportation, that are categorized as truck load (TL) and less than truck load (LTL) transportation. In Truck load transportation, the cost is a fixed for one truck up to a given capacity. However in some cases the weighted quantity may not be large enough to substantiate the cost associated with a TL mode. In such situation, a LTL mode may be used. LTL may be defined as a shipment of weighted quantity which does not fill a truck, and the transportation cost is taken on the basis of per unit weight. The cost – benefit measurements are very important in any business, the same is the case in the current study that retail store ABC desires to minimize the cost incurred on procurement and distribution supply chain.

The organization of the paper is as follows. Section 2 presents the details of model's assumptions, sets, and symbols and Section 3 provides the problem formulation, its analysis and price breaks along with the solution to the above discussed case. Section 4 gives conclusions of the current study and future directions of research in the context of the present study.

2. Model's Assumptions, Sets, and Symbols

The formulated mathematical model is based on the following assumptions, sets, and symbols.

2.1. Assumptions

The basic model makes the following assumptions:

1. Demand is deterministic.
2. Shortages are not allowed.
3. There is specific initial inventory at the beginning of planning horizon with supplier.
4. Two modes of transportation are used in second stage of transportation namely, truckload (TL) and less than truckload (LTL).

2.2. Sets

- Product set with cardinality P and indexed by i.
- Period set with cardinality T and indexed by t.
- Item discount break point set with cardinality L and indexed by small l.
- Freight discount break point set with cardinality K and indexed by small k.
- Waiting time set at intermediate stoppage with cardinality Γ and indexed by small τ.

2.3. Parameters

C Total cost

ϕ_{it} Unit purchase cost for i^{th} item in t^{th} period

d_{ilt} It reflects the fraction of regular price that the buyer pays for purchased items.

c_t Cost of unit weighted quantity of period t

f_{kt} It reflects the fraction of regular price that the buyer pays for transported weights.

O_t^τ Holding cost incurred on the total weight in period t for τ days

h_i Inventory holding cost per unit of item i

s Cost/kg of transportation in LTL policy

β_t Fixed freight cost for each truck load in period t

D_{it} Demand for item i in period t

IN_i Inventory level at the beginning of planning horizon for product i

w_i Per unit weight of item i

ω Weight transported in each full truck (in kgs.)

a_{ilt} Limit beyond which a price break becomes valid in period t for item i for l^{th} price break

b_{kt} Limit beyond which a freight break becomes valid at period t for k^{th} price break

2.4. Decision Variables

X_{it} Amount of item i ordered in period t

R_{ilt} If the ordered quantity falls in l^{th} price break then the variable takes value 1 otherwise zero

$$R_{ilt} = \begin{cases} 1 & \text{if } X_{it} \text{ falls in } l^{th} \text{ pricebreak} \\ 0 & \text{otherwise} \end{cases}$$

Z_{kt} If the weighted quantity transported falls in k^{th} price break then the variable takes value 1 otherwise zero

$$Z_{kt} = \begin{cases} 1 & \text{if } L_t \text{ falls in k pricebreak} \\ 0 & \text{otherwise} \end{cases}$$

L_{1t} Total weighted quantity transported in stage 1 of period t

$V_{\tau t}$ If the weighted quantity transported waits in t^{th} period for τ number of days then the variable takes value 1 otherwise zero

$$v_{\tau t} = \begin{cases} 1 & \text{if } L_t \text{ waits at halt} \\ 0 & \text{otherwise} \end{cases}$$

I_{it} Inventory level at the end of period t for product i

j_t Total number of truck loads in period t

y_t Amount in excess of truckload capacity (in weights)

u_t The variable u_t (or, $1-u_t$) reflects usage of policies, either TL and LTL policies or only TL policy i.e.

$$u_t = \begin{cases} 1 & \text{if considering TL \& LTL both policies} \\ 0 & \text{if considering only TL policy} \end{cases}$$

L_{2t} Total weighted quantity transported in stage 2 of period t.

3. Problem Formulation

The optimization problem minimizing the sum of total cost incurred in purchasing, holding and transportation subject to procurement and distribution constraints is formulated as follows.

3.1. *Mathematical Model Formulation*

Minimize

$$C = \sum_{t=1}^{6}\left[\left\{\sum_{i=1}^{5}\left(\sum_{l=1}^{4} R_{ilt}d_{ilt}\phi_{it}X_{it}\right)\right\} + \sum_{k=1}^{3} z_{kt}f_{kt}c_{t}L_{1t} + \sum_{\tau=1}^{3} L_{1t}O_t^{\tau}v_{\tau t}\right]$$

$$+ \sum_{t=1}^{6}\sum_{i=1}^{5} h_i I_{it} + \sum_{t=2}^{7}(\text{sy}_t + j_t\beta_t)u_t + (j_t+1)\beta_t(1\text{-}u_t) \tag{1}$$

subject to

$$I_{it} = I_{it-1} + X_{it} - D_{it} \quad \text{where } i=1,...,5;\ t=2,...,6 \tag{2}$$

$$I_{i1} = IN_{i1} + X_{i1} - D_{i1} \quad \text{where } i=1,...,5 \tag{3}$$

$$\sum_{t=1}^{6} I_{it} + \sum_{t=1}^{6} X_{it} \geq \sum_{t=1}^{6} D_{it} \quad \text{where } i=1,...,5 \tag{4}$$

$$X_{it} \geq \sum_{l=1}^{4} a_{ilt} R_{ilt} \quad i=1,...,5 \quad t=1,...,6 \tag{5}$$

$$\sum_{l=1}^{4} R_{ilt} = 1 \quad i=1,...,5,\ t=1,...,6 \tag{6}$$

$$L_{1t} = \sum_{i=1}^{5} w_i X_{it} \quad t = 1,...,6 \tag{7}$$

$$L_{1t} \ge \sum_{k=1}^{3} b_{kt} Z_{kt} \quad t = 1,...,6 \tag{8}$$

$$\sum_{k=1}^{3} Z_{kt} = 1 \quad t = 1,...,6 \tag{9}$$

$$L_{1\,t} = L_{2\,t+1} \quad t = 1,...,6 \tag{10}$$

$$L_{2t} \le (y_t + j_t \omega) u_t + (j_t + 1) \omega (1 - u_t) \quad t = 2,...,7 \tag{11}$$

$$L_{2t} = (y_t + j_t \omega) \quad t = 2,...,7 \tag{12}$$

$$\sum_{\tau=1}^{3} v_{\tau t} = 1 \quad t = 1,...,6 \tag{13}$$

$X_{it}, L_{1t}, L_{2t}, I_{it}, y_t, j_t \ge 0$ *and integers*;

$R_{ilt} = 0 \ or \ 1; \ Z_{kt} = 0 \ or \ 1; \ v_{\tau t} = 0 \ or \ 1; \ u_t = 0 \ or \ 1.$

3.2. Analysis of Model Formulation

The objective function (1) of the optimization problem is to minimize the sum of total cost incurred in purchasing the goods reflected by the first term of the objective function; transportation cost from the source to the intermediate stoppage reflected by the second term; holding cost at the intermediate stoppage of the weighted quantity reflected by the third term; ending inventory carrying cost at the source reflected by the fourth term and the last term reflects the transportation cost from intermediate stoppage to the final destination. The cost is calculated for the duration of the planning horizon.

Constraints (2 – 4) are the balancing equations, which calculate the ending inventory level in period t. In Eq. (2) ending inventory depends upon the inventory left in the last period, the quantity X_{it} ordered in period t and at demand D_{it}. Equation (3) calculates inventory level at the end of the first period for all the products using the inventory level at the beginning of the planning horizon, and the net change at the end of period one. Equation (4) takes care for shortages i.e. the sum of ending inventory and optimal order quantity is more than the demand of all the periods. Equation (5) shows that the order quantity of all items in period t exceeds the quantity break threshold. Equation (6) restricts the activation at exactly one level, either discount or no discount situation. The integrator for procurement and distribution is Equation (7), which calculates transported quantity according to item weight. Equations (8 – 9) work as Equations (5 – 6) with ordered quantity replaced by weighted quantity. Eq. (10) shows the total weighted quantity transported in stage 1 of period t is equal to the total weighted quantity transported in stage 2 of period $t + 1$. In Eq. (11), the minimum weighted quantity transported is calculated and further Eq. (12) measures the overhead units from truckload capacity in weights. From Eq. (13), we get to know the halting time.

Price Breaks

As discussed above, d_{ilt} variable specifies the fact that when the order size at period t is larger than a_{ilt} it results in discounted prices for the ordered items for which the price breaks are defined as:

Price breaks for ordering quantity are:

$$d_f = \begin{cases} d_{ilt} & a_{ilt} \leq X_{it} \leq a_{il+1t} \\ d_{iLt} & X_{it} \geq a_{iLt} \end{cases}$$

$$i = 1,...,5; \quad t = 1,...,6; \quad l = 1,...,4.$$

Freight breaks for transporting quantity are:

$$d_f = \begin{cases} f_{kt} & b_{kt} \leq L_t \leq b_{k+1t} \\ f_{kt} & L_t \geq b_{Kt} \end{cases}$$

$$k = 1,...,3; t = 1,...,6$$

where b_{kt} is the minimum required quantity to be transported in cargo.

3.3. *Solution*

The solution of the above formulated optimization problem is obtained by programming it in Lingo 11.0 software. LINGO is a comprehensive software tool designed to provide solutions of Linear, Nonlinear (convex & non-convex/Global), Quadratic, Quadratically Constrained, Integer optimization models in a fast and efficient manner. The required data sets parameters such as quantity demanded, various costs, initial inventory, weights per product, quantity thresholds and discounts tabulated in Appendix A (changed due to cutting edge competition and cannot be revealed but the model is applicable in same scenarios in big data) are fed in the program to generate the solution. In particular we have considered 6 periods and 5 products and attained the solution presented in Appendix B. In general we can incorporate any number of products and periods to obtain the solution of the problem.

The solution obtained and presented in Appendix B reflects that in first period, for all five electronic products considered in our case study above, the ordered quantity by retail store ABC is 110 units for DC, 305 units for LCD, 20 units for Laptop, 83 units of I-pod, and 205 units of AC with 6%, 15%, 0%, 4% and 10% discounts respectively in first stage of procurement. The ending inventory at source for period 1 is zero for all products. The weighted quantity transported in first stage is 6000 kgs for all products with discount of 8% on distribution of the same through Cargo. At halting point, the same weighted quantity will wait for two days on which first day is free. Now on the starting of 2nd period, distribution through trucks will start by using TL policy, with 6 full trucks. Similarly we can do the analysis of the given 5 electronic products in the periods 2 to 6.The total cost of the model is $7,667,683. Segregation of Costs at glance is shown in the following Table 1 below:

Table 1. Cost Parameters

	Holding Cost	Purchase Cost	Ist stage Transportation Cost	Halt Cost	IInd Stage Transportation Cost
Costs(in $)	460,024	4,561,347	1,314,262	1,275,350	56,700

4. Conclusions

In this paper we have investigated the two stage supply chain optimization model that minimizes purchasing, holding and transportation cost of electronic goods. The ordered quantity is transported from the single source to single destination through an intermediate stoppage at which the mode of transportation is changed. The different discount schemes are given by the supplier on the ordered quantity and by the transporter on the weighted quantity to the buyer. Finally the optimal ordered quantity, holding inventory and weighted transported quantity are determined. Hence we can conclude from our present research that integration of various functions of different entities is possible, in order to minimize the aggregate cost of purchasing and transportation activities. In fact the results of this study open several opportunities for further research and improvements. In future research, different extensions to the proposed models can be considered. Although the focus has been on single-stage models, we believe that these inventory models provide a strong foundation for subsequent analyses of multi-stage systems, since previous efforts in quantity discounts have primarily been focused on single-stage models involving only suppliers and producers.

Appendix A: Data Set

Table 2. Quantity Demanded of Digital Camera (DC), LCD, Laptop(LT), IPod, Air Conditioner (AC) (D_{it})

	Period 1	Period 2	Period 3	Period 4	Period 5	Period 6
DC	75	65	95	80	85	90
LCD	505	400	450	600	500	550
LT	60	170	150	200	80	100
IPod	75	65	95	80	60	60
AC	505	400	600	650	550	440

Table 3. Unit Cost (in $) per product per period (Φ_{it})

	Period 1	Period 2	Period 3	Period 4	Period 5	Period 6
DC	250	240	230	260	250	235
LCD	1290	1280	1270	1110	1285	1280
LT	645	535	725	600	745	650
IPod	250	240	230	250	260	240
AC	1290	1280	1270	1300	1290	440

Table 4. Maximum Initial inventory at the beginning of planning horizon (IN_{i1})

DC	LCD	LT	IPod	AC
50	256	40	60	300

Table 5. Transportation cost (in $) of weighted quantity from source to intermediate stoppage (c_t)

Period 1	Period 2	Period 3	Period 4	Period 5	Period 6
20	25	30	35	30	25

Table 6. Weight per product (w_i) per kg

DC	LCD	LT	IPod	AC
2	10	3	5	11

Table 7. Holding cost (in \$) at intermediate stoppage (o_t^τ)

Period 1	Period 2	Period 3	Period 4	Period 5	Period 6
15	20	25	35	30	30

Table 8. Fixed freight cost (in \$) for each truckload from intermediate stoppage to destination (β_t)

Period 1	Period 2	Period 3	Period 4	Period 5	Period 6
1000	1050	1100	1200	1000	1150

Table 9. Holding Cost (in \$) per product at destination (h_i)

DC	LCD	LT	IPod	AC
34	148	84	36	100

Table 10. Quantity thresholds (a_{ilt}) and discount factors (d_{ilt}) of Digital Camera (DC)

a_{ilt}	d_{ilt} Period 1,4	a_{ilt}	d_{ilt} Period 2,5	a_{ilt}	d_{ilt} Period 3,6
$0 \leq X_{it} < 70$	1	$0 \leq X_{it} < 75$	1	$0 \leq X_{it} < 80$	1
$70 \leq X_{it} < 80$	0.98	$75 \leq X_{it} < 85$	0.89	$80 \leq X_{it} < 90$	0.92
$80 \leq X_{it} < 90$	0.96	$85 \leq X_{it} < 95$	0.85	$90 \leq X_{it} < 100$	0.87
$90 \leq X_{it}$	0.94	$95 \leq X_{it}$	0.80	$100 \leq X_{it}$	0.82

Table 11. Quantity threshold (a_{ilt}) of LCD

a_{ilt}	d_{ilt} Period 1,4	a_{ilt}	d_{ilt} Period 2,5	a_{ilt}	d_{ilt} Period 3,6
$0 \leq X_{it} < 105$	1	$0 \leq X_{it} < 115$	1	$0 \leq X_{it} < 120$	1
$105 \leq X_{it} < 205$	0.95	$115 \leq X_{it} < 215$	0.85	$120 \leq X_{it} < 240$	0.79
$205 \leq X_{it} < 305$	0.90	$215 \leq X_{it} < 315$	0.82	$240 \leq X_{it} < 480$	0.72
$305 \leq X_{it}$	0.85	$315 \leq X_{it}$	0.75	$480 \leq X_{it}$	0.65

Table 12. Quantity threshold (a_{ilt}) of Laptop(LT)

a_{ilt}	d_{ilt} Period 1	a_{ilt}	d_{ilt} Period 2	a_{ilt}	d_{ilt} Period 3
$0 \leq X_{it} < 95$	1	$0 \leq X_{it} < 100$	1	$0 \leq X_{it} < 200$	1
$95 \leq X_{it} < 99$	0.75	$100 \leq X_{it} < 200$	0.87	$200 \leq X_{it} < 400$	0.82
$9 \leq X_{it} < 103$	0.70	$200 \leq X_{it} < 300$	0.83	$400 \leq X_{it} < 600$	0.73
$103 \leq X_{it}$	0.65	$300 \leq X_{it}$	0.78	$600 \leq X_{it}$	0.64

Table 13. Quantity threshold (a_{ilt}) of Ipod

a_{ilt}	d_{ilt} Period 1,4	a_{ilt}	d_{ilt} Period 2,5	a_{ilt}	d_{ilt} Period 3,6
$0 \leq X_{it} < 70$	1	$0 \leq X_{it} < 75$	1	$0 \leq X_{it} < 80$	1
$70 \leq X_{it} < 80$	0.98	$75 \leq X_{it} < 85$	0.89	$80 \leq X_{it} < 90$	0.92
$80 \leq X_{it} < 90$	0.96	$85 \leq X_{it} < 95$	0.85	$90 \leq X_{it} < 100$	0.87
$90 \leq X_{it}$	0.94	$95 \leq X_{it}$	0.80	$100 \leq X_{it}$	0.82

Table 14. Quantity threshold (a_{ilt}) of Air Conditioner (AC).

a_{ilt}	d_{ilt} Period 1,4	a_{ilt}	d_{ilt} Period 2,5	a_{ilt}	d_{ilt} Period 3,6
$0 \leq X_{it} < 105$	1	$0 \leq X_{it} < 115$	1	$0 \leq X_{it} < 120$	1
$105 \leq X_{it} < 205$	0.95	$115 \leq X_{it} < 215$	0.85	$120 \leq X_{it} < 240$	0.79
$205 \leq X_{it} < 305$	0.90	$215 \leq X_{it} < 315$	0.82	$240 \leq X_{it} < 480$	0.72
$305 \leq X_{it}$	0.85	$315 \leq X_{it}$	0.75	$480 \leq X_{it}$	0.65

Table 15. Weight thresholds (b_{kt}) and discount factors (f_{kt}) of transported weights

b_{kt}	f_{kt} Period 1,4	b_{kt}	f_{kt} Period 2,5	b_{kt}	f_{kt} Period 3,6
$300 \leq L_{lt} < 1000$	1	$375 \leq L_{lt} < 500$	1	$325 \leq L_{lt} < 750$	1
$1000 \leq L_{lt} < 2000$	0.96	$500 \leq L_{lt} < 1000$	0.98	$750 \leq L_{lt} < 1250$	0.92
$2000 \leq L_{lt} < 305$	0.92	$1000 \leq L_{lt} < 315$	0.94	$1250 \leq L_{lt} < 480$	0.88

Per Truckload capacity $\omega = 1000$; Cost per kg of transporting in LTL policy s= $2

Appendix B: Solution

Table 16. The Ordered Quantity (X_{it}) of given products in the respective periods

	Period 1	Period 2	Period 3	Period 4	Period 5	Period 6
DC	110	95	145	0	0	0
LCD	305	344	1234	316	0	0
LT	20	320	0	280	0	0
IPod	83	202	114	0	100	65
AC	205	400	1800	0	0	0

Table 17. Discounts availed by buyer on ordered quantity

	Period 1	Period 2	Period 3	Period 4	Period 5	Period 6
DC	6%	20%	13%	0%	0%	0%
LCD	15%	25%	35%	15%	0%	0%
LT	0%	22%	0%	35%	0%	0%
IPod	4%	20%	18%	0%	20%	0%
AC	10%	25%	35%	0%	0%	0%

Table 18. Transported weights in both the stages (L_{lt} & $L_{2\,t+1}$)

$L_{11} = L_{22}$	$L_{12} = L_{23}$	$L_{13} = L_{24}$	$L_{14} = L_{25}$	$L_{15} = L_{26}$	$L_{16} = L_{27}$
6000	10,000	33,000	4,000	500	325

Table 19. Discounts availed by buyer on weighted quantity:

Period 1	Period 2	Period 3	Period 4	Period 5	Period 6
8%	6%	12%	8%	2%	0%

Table 20. Waiting time at intermediate stoppage in days (τ)

$v_{\tau t}$	Day 1	Day 2	Day 3	Day 4
τ	Free	2	2	2

Table 21. Policy used for truckload schemes (u_t)

Period 2	Period 3	Period 4	Period 5	Period 6	Period 7
TL	TL & LTL	TL & LTL	TL & LTL	TL	LTL

Table 22. Number of trucks used for transportation (j_t)

Period 1	Period 2	Period 3	Period 4	Period 5	Period 6
6	10	33	4	1	0

Table 23. Overhead units (y_t)

Period 1	Period 2	Period 3	Period 4	Period 5	Period 6
0	0	0	0	500	325

Table 24. Ending Inventory at source (I_{it})

	Period 1	Period 2	Period 3	Period 4	Period 5	Period 6
DC	0	85	115	165	85	0
LCD	0	56	0	784	500	0
LT	0	0	150	0	80	0
IPod	0	8	145	164	84	124
AC	0	0	0	1200	550	0

References

1. H. S. Ahn and P. Kaminsky, *IIE Transactions* **37**, 609 (2005).
2. P. Kaminsky and D. S. Levi, *Production, Manufacturing and Logistics IIE Transactions* **35**, 1065 (2003).
3. B.Q. Rieksts and J.A. Ventura, *Computers & Operations Research* **37**, 20 (2010).
4. K.N. Subramanya and S.C. Sharma, *International Journal of Computer Science and Network Security* **9**, 364 (2009).
5. H. Wang, H. Liu and J. Yang, *Int. J. Adv. Manuf. Technol.* **43**, 200 (2009).

AN IMPROVED BOUND ON AVERAGE CODEWORD LENGTH OF VARIABLE LENGTH ERROR CORRECTING CODES

RICHA GUPTA* and BHU DEV SHARMA[†]

Jaypee Institute of Information Technology,
Noida, India
* E-mail: rgupta.ece@gmail.com
† E-mail: bhudev_sharma@yahoo.com

Reliable communication is an active area for research and applications. There is need to have efficient codes. Most codes considered have words of equal lengths, through algebraic tools. However, equal lengths idea compromises with efficiency which can be improved through variable length error correcting codes (VLECCs) for noisy channels. The area of developing VLECCs and mathematical tools for them has attracted quite some attention. Bernard and Sharma's[1,3] study provided a lead by initiating combinatorial approach for construction and existence of such codes by obtaining bounds on the average codeword length. This paper, extending that work, obtains an improvement over their bound. Some examples are included to elucidate this further.

1. Introduction

Early studies on coding theory involved two main directions - source coding (for noiseless channels) and channel coding (for noisy channels). Variable length codes were primarily employed for source coding. Kraft's[8] inequality was a significant bound on the codeword length of variable length source codes. It proved to be the necessary and sufficient condition for existence of a source code. The development of concepts of uniquely decodable codes and then instantaneously uniquely decodable codes led to the construction of the most efficient source codes by Huffman[7] that gave a method to construct prefix codes whose average codeword length was closest to the entropy, thereby maximising the efficiency. Shannon's[9,10] theorems for source coding and for existence of codes for reliable communication over unreliable channels provided a challenge for intensive research in development and studies in coding theory.

The in-built complexity of unique-decodability for variable length coding in case of noisy channels was by-passed by considering all code words of the same length. This additionally provided strong mathematical tools of algebra and finite fields for coding problems. Concept of parity checks and introduction of concept of distance by Hamming[5] provided another mathematical lever for defining / identifying errors in transmission. A variety of constant length error correcting codes like block codes, convolutional codes and cyclic codes enriched the area. In particular BCH codes, Reed-Solomon codes found wide applications. However, constant length coding compromised on efficiency. Developing variable length codes for noisy channels encountered two major difficulties that may briefly be identified as follows:

1. *Frame synchronization:* The asynchronous character of variable length codes requires the use of variable capacity storage buffers over a channel which limits the practical applications of these codes for channels which support data transmissions at fixed rates.

2. *Lack of mathematical tools:* With variable length words, linear algebraic tools are no longer of help. Even combinatorial results on bounds were not studied for a long time.

Some early thoughts on VLECCin this direction are available in Hartnett,[6] and a systematic study was undertaken by Bernard and Sharma[2,3] (see also Bernard[1]).

In this paper, Bernard and Sharma's study is extended further. Section 2 is preliminary, giving definitions and earlier results needed for our work. Section 3, carries the new result giving tighter bound on average codeword length for variable length error correcting codes. Two examples, with details in appendices, are given there to demonstrate the strength of the new improved bound derived. Section 4 briefly concludes the study indicating directions of continuing further work.

2. Preliminaries and Earlier Results

For error correction/detection, the basic idea is to find the set of possible strings of code characters that can be received when a given code-word is sent. For a code word, Hartnett[6] called this set 'error-admissibility range' of the code word. Based on this concept, Bernard and Sharma[3] derived necessary and sufficient conditions for the existence of variable length error correcting codes and by giving upper and lower combinatorial bounds on the lengths of code words, possessing specified error correction capabilities Interestingly, this study generalizes noiseless channel studies to noisy channels and creates an interface between constant-length and variable length error coding. We give some definitions and results that are used in this paper. We consider a code alphabet with q characters and let C denote a variable length code with K codewords $c_1, c_2, ..., c_K$ with respective word lengths $n_1, n_2, ..., n_K$.

Error-admissibility Range: The α error-admissibility mapping $\alpha(c)$ for a word c, is the set of words received over the channel when c is sent.

The mapping α in general will depend on the channel's error characteristics. For example, if the channel's error characteristics are random with at most e errors in a word c of length n, then this is given by

$$\alpha_e(c) = \{u | w(c - u) \le e\}$$

where w denotes the weight of the error-vector, that is difference of code vector c and received vector u.

Also, the size of this error-admissibility range is then given by

$$|\alpha_e|_n = \sum_{k=1}^{e} \binom{n}{k} (q - 1)^k$$

where, q is the size of the alphabet.

Extended Prefix Code: A code is said to be 'extended prefix code' if no vector in the error-range of a codeword is prefix to any vector in the error-range of other code words.

$\alpha-$Prompt Code: A code is called an $\alpha-$*prompt code* if it satisfies the extended prefix property under error-admissibility mapping.

Combinatorial Inequality on Word Lengths of Prompt Codes: Bernard and Sharma[2] generalized Kraft's inequality for $\alpha-$ prompt codes, based on some combinatorial results that is a necessary but not necessarily sufficient. It is stated in the following theorem.

Theorem 2.1 (Bernard and Sharma[2]). *An ($\alpha-$prompt Code) with K words of lengths $n_1 \le n_2 \le ... \le n_K$ over code alphabet of q symbols necessarily satisfies the condition:*

$$\sum_{i=1}^{K} |\alpha|_{n_i} q^{-n_i} \le 1, \tag{1}$$

where $|\alpha|_{n_i}$ denotes the size of the $\alpha-$range of a word of length n_i.

This result has been further generalised to the following situation called segment decomposition by Bernard and Sharma[3] as described below.

Segment decomposition: The idea of error range of a code word is further sharpened by considering what is termed 'segment decomposition.' The segment decomposition of the word c_i is obtained by putting the first n_1 entries in the first segment, $n_2 - n_1$ next entries in the second segment and the $n_i - n_{i-1}$ entries in the last segment. In this case the range $|\alpha|_n$ may be modified by the consideration of what is termed as the effective range as defined below.

Effective range: In terms of segment decomposition now we define the effective range as the product of the ranges of all the segments considering each segment as a different code word. Thus for the code as considered above, it is given by

$$|r_\alpha|_{n_i} = |\alpha|_{n_2 - n_1} \ldots |\alpha|_{n_i - n_{i-1}}.$$

This provides a *tighter bound* as an extension of Theorem 1 as given by:

Theorem 2.2 (Bernard and Sharma[2]). *An (α−prompt Code) with K words of lengths $n_1 \leq n_2 \leq \ldots \leq n_K$ over code alphabet of q symbols necessarily satisfies the condition:*

$$\sum_{i=1}^{K} |r_\alpha|_{n_i} q^{-n_i} \leq 1, \tag{2}$$

where $|r_\alpha|_{n_i}$ denotes the effective size of the α−range of a word of length n_i.

It may also be noted that it has been proved by Bernard and Sharma[3] that this condition is also a necessary condition for the existence of α−prompt code. Now In order to find a sufficient condition, we need a new mapping, which is defined next.

Detection Range: Given an error admissible mapping 'α' and a code C, mapping α^d is defined such that for any $c \in C$, an element of $\alpha^d(c)$ can be expressed as the difference of two elements of $\alpha(c)$. The mapping α^d is known as the 'detection range mapping of α' and the range of a codeword c under α^d will be called the detection range of c.

For $\alpha = \alpha(e)$ the detection range mapping α^d will simply be α_{2e}, *i.e.* the mapping such that for any $c \in C$, will consist of all vectors differing from c in up to $2e$ places.

We now state the theorem giving a sufficient condition for the existence of α−prompt code.

Theorem 2.3 (Bernard and Sharma[3]). *It is always possible to construct an $\alpha-$prompt code with K code words of lengths $n_1 \leq n_2 \leq ... \leq n_K$ if K is the smallest integer with these lengths satisfying*

$$\sum_{i=1}^{K} |\alpha^d|_{n_i} q^{-n_i} \geq 1. \tag{3}$$

The results of Theorems 2.1 and 2.2 reduce in respective situations to the Hamming[5] sphere packing bound when all code words are of equal length and to the necessary part of the Kraft's inequality, when the channel is noiseless. Also, the result of Theorem 2.3 reduces to the Gilbert[4] bound and the sufficient part of the Kraft's inequality under those situations. These results, therefore, give a new perspective on coding by linking the combinatorial bounds from the two different directions in which coding has developed. Next we present the result of Bernard and Sharma[3] on *a bound on average codeword length* for 'prompt codes.' As discussed earlier, these are the variable length error correcting codes that satisfy the 'extended prefix property'. The result is given below in the form of Theorem 2.4 (using necessary condition), an improvement of which will be provided next.

Theorem 2.4 (Bernard and Sharma[3]).
Considering a source S having K messages $m_1, m_2, ..., m_K$ with message probabilities $p_1, p_2, ..., p_K, \sum_{i=1}^{K} p_i = 1$. Let an $\alpha-$prompt code encode these messages into a code alphabet of q symbols and let the length of the codeword corresponding to message m_i be n_i. Then the average length $\bar{n}$ of the code satisfies

$$\bar{n} \geq \sum_{i=1}^{K} p_i \log \frac{|r_\alpha|_{n_i}}{p_i} \tag{4}$$

with equality if and only if, $n_i = \log \frac{|r_\alpha|_{n_i}}{p_i}$ for all i.

3. An Improved Bound on Average Codeword Length

In this section, we prove a lower bound on the average codeword length for $\alpha-$prompt codes, which is tighter than the one derived by Bernard and Sharma.[3]

Theorem 3.1. *Considering a source S having K messages $m_1, m_2, ..., m_K$ with message probabilities $p_1, p_2, ..., p_K, \sum_{i=1}^{K} = 1$. If an $\alpha-$prompt code encodes these messages using q alphabet symbols and let n_i denote* *the length of the codeword corresponding to message m_i, then the average length $\bar{n}$ of the code satisfies*

$$\bar{n} \geq \sum_{i=1}^{K} p_i \log_q \left(\frac{|\alpha^d|_{n_i}}{p_i} \right) - \sum_{i=1}^{K} p_i \log_q \left(\sum_{j=1}^{K} |\alpha^d|_{n_j} q^{-n_j} \right), \quad (5)$$

with equality if and only if, $p_i = \frac{|\alpha^d|_{n_i} q^{-n_i}}{\sum_{j=1}^{K} |\alpha^d|_{n_j} q^{-n_j}}$ *for all i.*

Proof: The approach we take is similar to that taken by Bernard and Sharma[3] in the proof of their bound. We have the well known inequality

$$-\sum_{i=1}^{K} p_i \log p_i \geq -\sum_{i=1}^{K} p_i \log q_i \quad (6)$$

for $p_i, q_i \geq 0$; $i = 1, 2, ..., K$ and $\sum_{i=1}^{K} p_i = \sum_{i=1}^{K} q_i = 1$, with equality iff $p_i = q_i$ for all i.

We take the probabilities q_i from the sufficient condition of the existence of variable length code-words, instead of taking it from the necessary condition of the existence of variable length code-words, *i.e.*

$$q_i = \frac{|\alpha^d|_{n_i} q^{-n_i}}{\sum_{j=1}^{K} |\alpha^d|_{n_j} q^{-n_j}}. \quad (7)$$

Then it follows from (6) that

$$-\sum_{i=1}^{K} p_i \log p_i \leq -\sum_{i=1}^{K} p_i \log(|\alpha^d|_{n_i} q^{-n_i}) + \sum_{i=1}^{K} p_i \log \left(\sum_{j=1}^{K} |\alpha^d|_{n_j} q^{-n_j} \right),$$

or

$$-\sum_{i=1}^{K} p_i \log p_i \leq \sum_{i=1}^{K} p_i n_i - \sum_{i=1}^{K} p_i \log(|\alpha^d|_{n_i} q^{-n_i}) + \sum_{i=1}^{K} p_i \log \left(\sum_{j=1}^{K} |\alpha^d|_{n_j} q^{-n_j} \right).$$

With $\bar{n} = \sum_{i=1}^{K} p_i n_i$, average code word length, and by Theorem 2.3, considering all logarithms to the base q, the above implies

$$\bar{n} \geq -\sum_{i=1}^{K} p_i \log_q p_i + \sum_{i=1}^{K} p_i \log_q(|\alpha^d|_{n_i} q^{-n_i}) - \sum_{i=1}^{K} p_i \log_q \left(\sum_{j=1}^{K} |\alpha^d|_{n_j} q^{-n_j} \right).$$

or

$$\bar{n} \geq \sum_{i=1}^{K} p_i \log_q(|\alpha^d|_{n_i} q^{-n_i}) - \sum_{i=1}^{K} p_i \log_q \left(\sum_{j=1}^{K} |\alpha^d|_{n_j} q^{-n_j} \right).$$

144

This proves the inequality in Eq. (5). The condition on equality follows based on the condition $p_i = q_i \forall i$, $i.e$

$$p_i = \frac{|\alpha^d|_{n_i} q^{-n_i}}{\sum_{j=1}^{K} |\alpha^d|_{n_j} q^{-n_j}}, \quad i = 1, 2, ..., K.$$

This complete the proof of the theorem.

$\square$

3.1. Examples

We consider two examples to explain the construction of binary ($q = 2$) variable length error correcting instantaneous codes and demonstrate the improvement withe new bound over the previous one. Let σ denote the number of different codeword lengths in the code C and let these lengths be $L_1, L_2, ..., L_\sigma$, where $L_1 < L_2 < ... < L_\sigma$. Let the number of code-words with length L_i be s_i. We shall use $(s_1@L_1, s_2@L_2, ...s_\sigma@L_\sigma)$ to denote such a code. In this set-up $n_1 = n_2 = ... = n_{s_1} = L_1; n_{s_1+1} = ... = n_{s_1+s_2} = L_2$ and so-on. The values of p_i are given in the corresponding tables in the appendices.

Example 1. This concerns a variable length error correcting instantaneous ($\alpha-$prompt) code C : $(1@4, s2@7)$, with two distinct lengths $L_1 = 4, L_2 = 7$ and $\alpha = \alpha_1$ (single error correcting). The range for the number of code words of length L_2 has been obtained using Theorem 2.2 and Theorem 2.3 (see Appendix A for calculation of the range of s_2). The obtained range is 2-4 (refer Appendix A). The basic lower bound on the average codeword length, $\bar{n}_n$, is given by Theorem 2.3. The improved lower bound on the average codeword length, $\bar{n}_s$, is given by Theorem 3.1. The actual average codeword length is represented by $\bar{n}$. Table 1 presents the comparison of $\bar{n}$ with $\bar{n}_n$ and $\bar{n}_s$ and for different values of s_2.

Table 1. Comparison of $\bar{n}_n$ and $\bar{n}_s$ with $\bar{n}$, for example 1.

	s_2		
	2	3	4
$\bar{n}$	5.50	5.80	5.80
$\bar{n}_s$	5.47	5.77	5.79
$\bar{n}_n$	4.82	5.44	5.69

It may be noted that the improved bound gives a tighter bound for $\bar{n}$ compared to $\bar{n}_n$.

Example 2. Here we are concerned with variable length code words of two different lengths $L_1 = 3, L_2 = 9$ and $\alpha = \alpha_1$ (single error correcting) and s_1 is taken as 1. Thus we want to construct code $C : (1@3, s_2@9)$. The range for s_2 has been obtained using Theorem 2.2 and Theorem 2.3 as detailed in Appendix B. It has been found that s_2 ranges from 2 to 9, however, it is possible to construct the code-words for s_2 from 2 to 6 only. The comparison of actual average codeword length and its two lower bounds ($\bar{n}_n$ and $\bar{n}_s$) is given in Table 2 below for several values of s_2.

Table 2. Comparison of $\bar{n}_n$ and $\bar{n}_s$ with $\bar{n}$, for example 2.

	s_2				
	2	3	4	5	6
$\bar{n}$	6.00	6.60	6.60	7.20	6.60
$\bar{n}_s$	5.58	6.08	6.29	6.74	6.45
$\bar{n}_n$	4.90	5.60	5.85	6.41	6.20

It can be seen that in all the cases, the improved bound gives a tighter lower bound on the average codeword length. The lists of code-words for the two examples considered here are given in Appendix A and Appendix B, respectively. Examples of single error correcting variable length codes have been considered here, although the theorems can be applied for any variable length codes of $e-$error correcting where e is a finite number.

4. Conclusion

In this paper, a new and tighter bound on the average code word length for α prompt code has been derived for variable length error correcting codes. This improved lower bound is much closer to the actual average codeword length as compared to the previously obtained bound. This result has been verified with different examples. The study of variable length error correcting codes is a rewarding area. We are engaged in developing algorithms for coding and decoding these codes.

Appendix A. The Range of s_2 for Example 1

The range for s_2 can be found by applying necessary and sufficient conditions for the existence of $\alpha-$prompt codes. First apply the necessary condition

$$\sum_{i=1}^{K} |r_\alpha|_{n_i} q^{-n_i} \leq 1.$$

We have $|r_\alpha|_4 = 5$ and $|r_\alpha|_7 = |r_\alpha|_4|r_\alpha|_3 = 5 \times 4 = 20$. Hence, s_2 from Theorem 2.2 satisfies

$$1 \times 5 \times 2^{-4} + s_2 \times 20 \times 2^{-7} \leq 1$$

that yields

$$s_2 \leq 4.4. \tag{A.1}$$

Now we use the sufficient condition from Theorem 2.3, namely

$$\sum_{i=1}^{K} |\alpha^d|_{n_i} q^{-n_i} \geq 1.$$

Here $|\alpha^d|_4 = |\alpha_2|_4 = 11$ and $|\alpha^d|_7 = |\alpha_2|_7 = 29$ and the sufficient condition implies

$$1 \times 11 \times 2^{-4} + s_2 \times 29 \times 2^{-7} \geq 1$$

that yields

$$s_2 \geq 1.37. \tag{A.2}$$

From equations (A.1) and (A.2) the range of s_2 is given by $2 \leq s_2 \leq 4$. The set of code-words for the same is listed in Table A1.

It can be seen in Table A1 that for each code no word in the range of any code word is prefix to any word in the range of any other codeword. So, the extended prefix property is satisfied.

Appendix B. The Range of s_2 for Example 2

Recall that in this example, we are considering the code $C : (1@3, s_2@9)$ with single error correcting code, that is, $L_1 = 3, L_2 = 9$ and $\alpha = \alpha_1$. In this example We have $|r_\alpha|_3 = 4$ and $|r_\alpha|_9 = |r_\alpha|_3|r_\alpha|_6 = 4 \times 7 = 28$. Hence, using the necessary condition from Theorem 2.2 s_2 satisfies

$$1 \times 4 \times 2^{-3} + s_2 \times 28 \times 2^{-9} \leq 1$$

that yields

$$s_2 \leq 9.14. \tag{B.1}$$

Now in order to use the the sufficient condition from Theorem 2.3, $|\alpha^d|_3 = |\alpha_2|_3 = 7$ and $|\alpha^d|_9 = |\alpha_2|_9 = 46$ and that implies

$$1 \times 7 \times 2^{-3} + s_2 \times 46 \times 2^{-9} \geq 1$$

Table A1.   Possible set of code words along with their ranges for $L_1 = 4, L_2 = 7$ and $\alpha = \alpha_1$.

s_1	s_2	Code words	$\alpha-$ admissibility mapping range
1	2	$c_1 = \{0000\}$ $p_1 = 0.5$	$r_\alpha(c_1) = \{0000, 0001, 0010, 0100, 1000\}$
		$c_2 = \{1111000\}$ $p_2 = 0.25$	$r_\alpha(c_2) = \{1111000, 1111001, 1111010, 1111001,$ $1110000, 1101000, 1011000, 0111000\}$
		$c_3 = \{0111111\}$ $p_3 = 0.25$	$r_\alpha(c_3) = \{0111111, 0111110, 0111101, 0111011,$ $0110111, 0101111, 0011111, 1111111\}$
1	3	$c_1 = \{0000\}$ $p_1 = 0.4$	$r_\alpha(c_1) = \{0000, 0001, 0010, 0100, 1000\}$
		$c_2 = \{1111000\}$ $p_2 = 0.2$	$r_\alpha(c_2) = \{1111000, 1111001, 1111010, 1111001,$ $1110000, 1101000, 1011000, 0111000\}$
		$c_3 = \{0111111\}$ $p_3 = 0.2$	$r_\alpha(c_3) = \{0111111, 0111110, 0111101, 0111011,$ $0110111, 0101111, 0011111, 1111111\}$
		$c_4 = \{1110101\}$ $p_4 = 0.2$	$r_\alpha(c_4) = \{1110101, 1110100, 1110111, 1110001,$ $1111101, 1100101, 1010101, 0110101\}$
1	4	$c_1 = \{0000\}$ $p_1 = 0.4$	$r_\alpha(c_1) = \{0000, 0001, 0010, 0100, 1000\}$
		$c_2 = \{1111000\}$ $p_2 = 0.2$	$r_\alpha(c_2) = \{1111000, 1111001, 1111010, 1111001,$ $1110000, 1101000, 1011000, 0111000\}$
		$c_3 = \{0111111\}$ $p_3 = 0.2$	$r_\alpha(c_3) = \{0111111, 0111110, 0111101, 0111011,$ $0110111, 0101111, 0011111, 1111111\}$
		$c_4 = \{1110101\}$ $p_4 = 0.2$	$r_\alpha(c_4) = \{1110101, 1110100, 1110111, 1110001,$ $1111101, 1100101, 1010101, 0110101\}$
		$c_5 = \{1011011\}$ $p_5 = 0.2$	$r_\alpha(c_5) = \{1011011, 1011010, 1011001, 1011111,$ $1010011, 1001011, 1111011, 0011011\}$

that yields

$$s_2 \geq 1.39. \tag{B.2}$$

From equations (B.1) and (B.2) the range of s_2 is given by $2 \leq s_2 \leq 9$. However, the codewords for $s_2 = 7, 8, 9$ are not possible. The set of codewords for the same is listed in Table B1 below. The extended prefix property is satisfied in this example also.

148

Table B1. Possible set of code words along with their ranges for $L_1 = 4, L_2 = 7$ and $\alpha = \alpha_1$.

s_1	s_2	Code words	$\alpha-$ admissibility mapping range
1	2	$c_1 = \{000\}$	$r_\alpha(c_1) = \{000, 001, 010, 1000\}$
		$p_1 = 0.5$	
		$c_2 = \{111000000\}$	$r_\alpha(c_2) = \{111000000, 111000001, 111000010, 111000100, 111001000,$
		$p_2 = 0.25$	$111010000, 110000000, 101000000, 011000000, 111100000\}$
		$c_3 = \{111000111\}$	$r_\alpha(c_3) = \{111000111, 111000110, 111000101, 111000011, 111001111,$
		$p_3 = 0.25$	$111010111, 111100111, 110000111, 101000111, 011000111\}$
1	3	$c_1 = \{000\}$	$r_\alpha(c_1) = \{000, 001, 010, 1000\}$
		$p_1 = 0.4$	
		$c_2 = \{111000000\}$	$r_\alpha(c_2) = \{111000000, 111000001, 111000010, 111000100, 111001000,$
		$p_2 = 0.2$	$111010000, 110000000, 101000000, 011000000, 111100000\}$
		$c_3 = \{111000111\}$	$r_\alpha(c_3) = \{111000111, 111000110, 111000101, 111000011, 111001111,$
		$p_3 = 0.2$	$111010111, 111100111, 110000111, 101000111, 011000111\}$
		$c_4 = \{111111000\}$	$r_\alpha(c_4) = \{111111000, 111111001, 111111010, 111111100, 111110000,$
		$p_4 = 0.2$	$111101000, 111011000, 110111000, 101111000, 011111000\}$
1	4	$c_1 = \{000\}$	$r_\alpha(c_1) = \{000, 001, 010, 1000\}$
		$p_1 = 0.4$	
		$c_2 = \{111000000\}$	$r_\alpha(c_2) = \{111000000, 111000001, 111000010, 111000100, 111001000,$
		$p_2 = 0.15$	$111010000, 110000000, 101000000, 011000000, 111100000\}$
		$c_3 = \{111000111\}$	$r_\alpha(c_3) = \{111000111, 111000110, 111000101, 111000011, 111001111,$
		$p_3 = 0.15$	$111010111, 111100111, 110000111, 101000111, 011000111\}$
		$c_4 = \{111111000\}$	$r_\alpha(c_4) = \{111111000, 111111001, 111111010, 111111100, 111110000,$
		$p_4 = 0.15$	$111101000, 111011000, 110111000, 101111000, 011111000\}$
		$c_5 = \{111111000\}$	$r_\alpha(c_5) = \{111110011, 111110010, 111110001, 111110111, 111111011,$
		$p_5 = 0.15$	$111100011, 111010011, 110110011, 101110011, 011110011\}$
1	5	$c_1 = \{000\}$	$r_\alpha(c_1) = \{000, 001, 010, 1000\}$
		$p_1 = 0.3$	
		$c_2 = \{111000000\}$	$r_\alpha(c_2) = \{111000000, 111000001, 111000010, 111000100, 111001000,$
		$p_2 = 0.2$	$111010000, 110000000, 101000000, 011000000, 111100000\}$
		$c_3 = \{111000111\}$	$r_\alpha(c_3) = \{111000111, 111000110, 111000101, 111000011, 111001111,$
		$p_3 = 0.2$	$111010111, 111100111, 110000111, 101000111, 011000111\}$
		$c_4 = \{111111000\}$	$r_\alpha(c_4) = \{111111000, 111111001, 111111010, 111111100, 111110000,$
		$p_4 = 0.1$	$111101000, 111011000, 110111000, 101111000, 011111000\}$
		$c_5 = \{111111000\}$	$r_\alpha(c_5) = \{111110011, 111110010, 111110001, 111110111, 111111011,$
		$p_5 = 0.1$	$111100011, 111010011, 110110011, 101110011, 011110011\}$
		$c_6 = \{111111000\}$	$r_\alpha(c_6) = \{111101110, 111101111, 111101100, 111101010, 111100110,$
		$p_6 = 0.1$	$111111110, 111001110, 110101110, 101101110, 011101110\}$
1	6	$c_1 = \{000\}$	$r_\alpha(c_1) = \{000, 001, 010, 1000\}$
		$p_1 = 0.4$	
		$c_2 = \{111000000\}$	$r_\alpha(c_2) = \{111000000, 111000001, 111000010, 111000100, 111001000,$
		$p_2 = 0.1$	$111010000, 110000000, 101000000, 011000000, 111100000\}$
		$c_3 = \{111000111\}$	$r_\alpha(c_3) = \{111000111, 111000110, 111000101, 111000011, 111001111,$
		$p_3 = 0.1$	$111010111, 111100111, 110000111, 101000111, 011000111\}$
		$c_4 = \{111111000\}$	$r_\alpha(c_4) = \{111111000, 111111001, 111111010, 111111100, 111110000,$
		$p_4 = 0.1$	$111101000, 111011000, 110111000, 101111000, 011111000\}$
		$c_5 = \{111111000\}$	$r_\alpha(c_5) = \{111110011, 111110010, 111110001, 111110111, 111111011,$
		$p_5 = 0.1$	$111100011, 111010011, 110110011, 101110011, 011110011\}$
		$c_6 = \{111111000\}$	$r_\alpha(c_6) = \{111101110, 111101111, 111101100, 111101010, 111100110,$
		$p_6 = 0.1$	$111111110, 111001110, 110101110, 101101110, 011101110\}$
		$c_7 = \{111111000\}$	$r_\alpha(c_7) = \{111101101, 111101100, 111101111, 111101001, 111100101,$
		$p_7 = 0.1$	$111111101, 111001101, 110101101, 101101101, 011101101\}$

References

1. M. Bernard , *Error Correcting Codes with Variable Lengths and Non- Uniform Errors*, PhD. Thesis, Department of Mathematics, Faculty of Natural Sciences, University of The West Indies, St. Augustine, Trinidad, West Indies (1987).

2. M. Bernard and Bhu Dev Sharma, *ARS Combinatoria* **25 B**, 181 (1988).
3. M. Bernard and Bhu Dev Sharma, *IEEE Trans. Information Theory* **36**, 1474 (1990).
4. E. N. Gilbert and E. F. Moore, *Bell Sys. Tech. J.* **38**, 933 (1959).
5. R. W. Hamming, , *Bell Sys. Tech. J.* **29**, 147 (1950).
6. W. E. Hartnett (Ed.), *Foundation of Coding Theory* (D. Reidel Publishing Co., Dordrecht, Holland, 1974).
7. D. Huffman, *Proceedings of IRE* **40**, 1098 (1962).
8. L. Kraft, *A Device for Quantizing, Grouping, and Coding Amplitude Modulated Pulses*, M.S. Thesis , Dept. of Elec. Eng., MIT, Cambridge, MA (1949).
9. C. E.Shannon, *Bell System Technical Journal* **27**, 379(1948).
10. C. E.Shannon, *Bell System Technical Journal* **27**, 623(1948).

TRIMMED ANALYSIS OF VARIANCE: A ROBUST MODIFICATION OF ANOVA

GOVIND S. MUDHOLKAR

Department of Statistics, University of Rochester
Rochester, NY 14623, USA

DEO KUMAR SRIVASTAVA

Department of Biostatistics, St. Jude Children's Research Hospital
Memphis, TN 38105, USA

CAROL E. MARCHETTI

Department of Mathematics and Statistics, Rochester Institute of Technology,
Rochester, NY 14623, USA

ANIL G. MUDHOLKAR

On Shore Networks LLC, 1407 West Chicago Avenue
Chicago, IL 60642, USA

The analysis of linear models in general and the analysis of variance in particular are among the most commonly used statistical techniques in bio-medical, engineering and social sciences. Introduced by R.A. Fisher nearly three quarters of a century ago, the analysis of variance, especially in its simplest one-way classification form, remains a workhorse in applied research to date. Its current importance is highlighted by Gelman[10] in the essay entitled "Analysis of Variance – Why it is more important than ever". The normal theory analysis of variance is easy to motivate, simple to implement and convenient for post-hoc analysis based upon simultaneous confidence intervals and multiple comparisons. Yet, the basic assumptions such as normality and homoscedasticity, which make the ANOVA practically attractive, have always been suspect. The objective of this work is to develop a robust version of the usual variance-ratio statistic by replacing the means by trimmed means and using appropriate studentization. A simple approximation for its null-distribution in terms of the F-distribution makes the statistic very practical and provides excellent type I error control across a broad spectrum of populations and for moderate sample sizes and permits the post-hoc analysis as in the normal theory case. An extensive Monte Carlo study shows that the robust test involves minor power loss in case when the normality assumption holds, but results in substantial power gain in case of heavy-tailed non-normal populations.

1. Introduction

The analysis of variance especially in the one-way classification setting is among the most commonly employed statistical tools in sciences and technology. This is in part because of the simplicity of its motivation and implementation and the well-established ANOVA table used for summarizing and communicating its results. Its basic normal theory version has historically been central in statistical education. The post-hoc analysis tools such as multiple comparisons and simultaneous confidence intervals associated with it make it more relevant in applied setting and widely available software packages such as SAS, S-Plus, R and SPSS have made its use easier. Yet, the assumptions such as normality, homoscedasticity often go unattended. Hence methods which are "resistant," as termed by Mosteller and Tukey[21], to minor departures from the assumptions or "robust," in the more common terminology, become relevant.

The best known solutions to the violation of homoscedasticity in one-way ANOVA are due to James[16], Welch[39] and Brown and Forsythe[5]; see also Hsu[13] and Lawton[18]. The normality violation, but assuming equal scale parameters in the location-scale framework, for testing homogeneity of the location parameters was addressed by Kruskal and Wallis[17] by offering their well-known rank test. It may be noted that the nonparametric tests, in general, are large sample procedures and not very amenable to post-hoc analysis. A more generally applicable data-transformation approach for analyzing data in linear models was proposed by Box and Cox[3]. In this paper we focus on modifying the familiar variance ratio statistic, which can be used as in ANOVA setting. However, it is robust, in the sense that it is reasonable in terms of the type I error control in a wide spectrum of data and offers substantial power advantages.

The shift in statistical thinking from nonparametric methods to the current robust methods can be said to be initiated by Tukey's[36] seminal article on the occurrence of heavier than normal tails and empirical study of the trimmed means. Also noteworthy is Tukey's[37] influential article on *future of data analysis*. The work on robust inference after that focused on studying of three basic types, the L- M- and R-estimators of location culminating in the famous collaborative Princeton Study published by Andrews *et al.*[1] Excellent accounts of the robustness theory, with emphasis mainly on estimation of location parameters, may be found in monographs such as Huber[15] and Hampel *et al.*[12]

The earliest work on robust testing and confidence interval estimation is by Tukey and Mclaughlin[38] where they constructed a robust analogue of one sample t-test by "trial and error" and using a "curious pairing," as noted by Huber[14], for studentizing trimmed means using Winsorized standard deviation. Huber[14] validates the result by appealing to the asymptotic distribution of trimmed mean, but we note that the same result can be established by appealing to the asymptotic theory of jackknife standard errors; e.g. see Thorburn[35]. Soon followed the extension of Tukey and McLaughlin's work by Yuen and Dixon[42] for comparing two means in homoscedastic case and by Yuen[41] in the unequal variances case. Later, Mudholkar *et al.*[22] presented a construction of a robust trimmed pooled-t statistic and empirically found its performance to be excellent in samples as small as 7 each. They also proposed an asymptotically valid version of the robust two sample t-statistic in heteroscedastic case. Around the same time there was parallel progress in development of one- and two-sample tests based on quick estimators such as the Trimean and Gastwirth estimator; see Mudholkar and Patel[23], Patel *et al.*[24], and Srivastava *et al.*[31]

In the literature there exist a few proposals for analysis of variance which are robust to violations of the normality assumption. These are essentially extensions of such methods for the two sample robust methods based on the L-, M- and R-estimators of location. For example, Schrader and Hettmansperger[27] consider likelihood ratio type tests involving *M*-estimators whereas Giloni, Seshadri and Simonoff[11] propose the use of Tukey's bisquare function to achieve a balance between robustness and efficiency. Bertaccini and Varriale[2] propose an approach based on Forward Search to detect and investigate the effect of observations that differ from the bulk of the data. However, even for moderate contamination proportions (in 10% range) their approach is unable to maintain type I error control and seems to worsen with increasing sample sizes. Lee and Fung[19] present statistics using trimmed means and sine-wave *M*-estimators. Other references to robust tests and simultaneous confidence intervals include Schrader and McKean[28], Tan and Tabatabai[33,34], Dunnett[7], Fung and Tam[8], Garcia-Perez.[9] A good discussion and comparison of such methods with respect to robustness and usability in the context of linear models can be found in Draper.[6]

In general, the robustness research based on the use of trimmed means assumes the populations to be symmetric and heavy tailed and use symmetric trimming of the tails in light of the justification of their asymptotic normality, e.g., see Stigler[32], Shorack[29,30]. However, it may be noted that, in case of distributions with tails lighter than normal require trimming inliers, e.g. see Shorack[29] and Prescott[25], i.e. the tests need to be based on the outside trimmings. Wilcox *et al.*[40], Luh and Guo[20], on the other hand, adopt the James and Welch statistic, substitute trimmed means and Winsorized variances in place of mean and sample variances and propose the robust versions of the F-test. Based on the simulation studies their recommendation is to use bootstrap approach as it provides reasonable operating characteristics. This may be said of the bootstrap approach in general. Both studies mentioned above adopt Bradley's[4] criterion which defines an (α) test to be robust only its empirical type I error is in the interval $0.5\alpha \leq \hat{\alpha} \leq 1.5\alpha$. They empirically examine the performance of the tests for various symmetric as well as asymmetric populations and symmetrically trimming the two tails. This approach may be justified in light of Winsor's principle, which suggests that, in practice, the distributions of very large samples are reasonably "normal in the middle."

In this paper we present a robust $\tilde{F}$-statistic for the homoscedastic case which uses trimmed means in place of sample means and pooled estimate of the common scale parameter based on Winsorized sample variances. This approach is easy to motivate and implement, is amenable to post-hoc analysis in the same manner as the classical F-test. An extensive Monte Carlo study shows that it is robust to departures from the normality assumption and can be used as in the normal theory ANOVA. This approach can be easily extended to the heteroscedastic case which results in test statistic that is asymptotically valid even when the underlying population is not normal.

The background of trimmed means is provided in Section 2 and the construction of the robust version of analysis of variance F-test developed in two stages is described in Section 2.1. The operating characteristics of the test are discussed in Section 3. Finally, Section 4 is devoted to conclusions and miscellaneous remarks.

2. Trimmed-Means ANOVA

The analysis of variance based on sample means is variously optimal, e.g. using invariance argument, Bayesian reasoning or maximin criterion, if the normality assumption holds. However, in case of heavier tailed populations a modification based on the use of trimmed means in place of the means can be justified in light of Tukey's[36] article or influence functions as in Hampel *et al.*[12] or Huber[15].

Trimmed Means. Let $X_1 < X_2 < \ldots < X_n$ be the order statistics of a random sample from a location scale population with symmetric distribution function $F\left((x-\theta)/\sigma\right)$. For an integer $g < n$, the δ-trimmed mean, $\delta = g/n$, is given by,

$$\tilde{X} = (X_{g+1} + \ldots + X_{n-g})/(n-2g).$$ (1)

The earliest use of the trimmed mean for developing a robust alternative to the one-sample Student's t-statistic was by Tukey and McLaughlin[38]. They proposed studentizing the trimmed mean by the Winsorized variance given by

$$\tilde{s}^2 = SS_W/h(h-1),$$ (2)

where $h = (n-2g)$ and SS_W is the Winsorized sum of squares given by,

$$SS_W = [(g+1)(X_{g+1} - \tilde{X})^2 + (X_{g+2} - \tilde{X})^2 + \ldots + (g+1)(X_{n-g} - \tilde{X})^2],$$ (3)

and constructed the trimmed t-statistic, as a robust substitute for the Student's t as

$$\tilde{t} = \frac{\tilde{X} - \theta}{\tilde{s}} . \tag{4}$$

Furthermore, they suggested approximating its null distribution by a Student's t-distribution with $(h-1)$ degrees of freedom.

Huber[14] justified the studentization in light of the asymptotic normal distribution of the trimmed means. Specifically, he showed that when the underlying distribution F is symmetric, continuous with mean θ and variance σ^2, and strictly increasing at points $\pm\xi$, then, asymptotically as $n \to \infty$,

$$\sqrt{n}(\tilde{X} - \theta) \rightarrow N(0, \sigma^2(\delta)), \tag{5}$$

where $\delta = F(-\xi)$ is the limit of the fraction g / n, and

$$\sigma^2(\delta) = \left[\int_{-\xi}^{\xi} x^2 \, dF + 2\delta \, \xi^2 \right] / (1 - 2\delta)^2$$

$$= V^2(\delta) \, \sigma^2, \quad \text{say} \tag{6}$$

Huber also showed that, as $n \to \infty$,

$$\sqrt{n-1} \left(\tilde{s}^2 - V^2(\delta) \, \sigma^2 \right) \rightarrow N\left(0, U^2(\delta) \, \sigma^4 \right), \tag{7}$$

where $\tilde{s}^2 = SS_W / n(1 - 2\delta)^2$, and SS_W is as given in (3) and

$$U^2(\delta)(1 - 2\delta)^4 \, \sigma^4 = \int_{-\xi}^{\xi} x^4 \, dF + 2\delta \left(\xi^2 + \frac{2\delta\xi}{f(\xi)} \right)^2 \tag{8}$$

$$- \left[\int_{-\xi}^{\xi} x^2 \, dF + 2\delta \left(\xi^2 + \frac{2\delta\xi}{f(\xi)} \right) \right]^2 .$$

However, after validating the Tukey-McLaughlin proposal, he emphasized the importance of carefully evaluating the properties of $\tilde{t}$. Yuen and Dixon[42] extended Tukey and McLaughlin's test to two homoscedastic samples with equal sample sizes and proposed to approximate its null distribution by a Student's t-distribution with $2(h-1)$ degrees of freedom.

Pooled Trimmed t-test. Mudholkar, Mudholkar and Srivastava[22] note that if the sample sizes are unequal then the above approximation is in error, even asymptotically, and propose an alternative construction of the pooled trimmed -t statistic. They begin by fixing the underlying $d.f.$ F at the normal $d.f.$ Φ and developing an approximation to the asymptotic distributions by starting as follows:

$$\sqrt{n}\left(\tilde{X} - \theta\right) \rightarrow N\left(0, V_{\Phi}^2(\delta)\sigma^2\right) \tag{9}$$

and

$$\sqrt{n-1}\left(\tilde{s}^2 - V_{\Phi}^2(\delta)\sigma^2\right) \rightarrow N\left(0, U_{\Phi}^2(\delta)\sigma^4\right). \tag{10}$$

Then, using a Monte Carlo experiment they empirically obtain,

$$V_{\Phi}^2(\delta) \approx 1 + 0.48\delta + 1.21\delta^2 \tag{11}$$

and

$$w_{\Phi} = (n-1)V_{\Phi}^4 / U_{\Phi}^2 \approx (n-1)\left(0.5 - 1.62\delta + 1.91\delta^2 - 1.85\delta^3\right). \tag{12}$$

Now, let the two ordered samples $X_{1,1} < X_{1,2} < ... < X_{1,n_1}$ and $X_{2,1} < X_{2,2} < ... < X_{2,n_2}$ be from two symmetric populations with means θ_1 and θ_2, respectively, and common variance σ^2. Let $\tilde{X}_1$, and $\tilde{X}_2$ be the δ_1- and δ_2-trimmed means, respectively, and let $\tilde{s}_1^2$ and $\tilde{s}_2^2$ be the corresponding Winsorized variances. Then, the pooled estimator of the common variance σ^2 can be obtained as the weighted mean of $\tilde{s}_1^2$ and $\tilde{s}_2^2$ with weights inversely proportional to their asymptotic variances, as in (10), given by,

$$\tilde{s}_p^2 = \left[w_1\left(\tilde{s}_1^2 / V_1^2\right) + w_2\left(\tilde{s}_2^2 / V_2^2\right)\right]\left(w_1 + w_2\right)^{-1} \tag{13}$$

where $V_i^2 = V_{\Phi_i}^2$, and $w_i = w_{\Phi_i}$ $i = 1, 2$ are given by equations (11) and (12). Note that in case of normal populations and no trimming (13) reduces to the familiar pooled estimator of the variance.

Then, the robust pooled trimmed-t statistic for testing the null hypothesis regarding θ_1 and θ_2 is given by,

$$\tilde{t}_p = \frac{(\tilde{X}_1 - \tilde{X}_2) - (\theta_1 - \theta_2)}{\tilde{s}_p\sqrt{V_1^2 / n_1 + V_2^2 / n_2}}. \tag{14}$$

An approximation to the null distribution of $\tilde{t}_p$, using a Monte Carlo experiment and the asymptotic theory, obtained in terms of a scaled Student's t -variate,

$$\tilde{t}_p \quad \sim \quad At_v, \tag{15}$$

where $\qquad\qquad v = 2\left(w_1 + w_2\right)$

and $\qquad\qquad A = 1 - 1.3\dfrac{\delta}{v} + 7.5\dfrac{\delta^2}{v} + 16\dfrac{\delta}{v^2} - 150\dfrac{\delta^3}{v^2},$

where w_i, $i = 1, 2$ are given by (12) and $\delta = (\delta_1 + \delta_2)/2$.

An empirical evaluation showed that the pooled trimmed t-statistic in (14) provides excellent type I error control and significant gain in power for heavy tailed distribution even for samples as small as 7. However, this empirical approach to adjusting the asymptotic distribution for use in moderate size samples is impractical for implementation in the ANOVA cases. We propose an alternative that makes the adjustments to the asymptotic distributions in (9) and (10) and develop the robust analog, $\tilde{F}$, of the normal theory F-statistic.

2.1. The Robust $\tilde{F}$ -statistic: Construction

The construction of a robust ANOVA type F -test based upon the trimmed means $(\tilde{X}_i)$ and Wisorized variances $(\tilde{s}_i^2)$ is now presented in two steps. First, a preliminary test statistic $\tilde{F}_{pre}$ based on $\tilde{X}_i$ and $\tilde{s}_i^2$ is developed, by analogy with the normal case, to have asymptotically correct null distribution. This null distribution of $\tilde{F}_{pre}$ is then empirically adjusted to obtain $\tilde{F}$ ANOVA statistic, which is useable with small to moderate size samples.

Let $X_{ij}, j = 1, 2, \ldots, n_i$ and $i = 1, 2, \ldots, r$ be random samples from symmetric location-scale distributions $F((X - \theta_i)/\sigma)$ with the common scale parameter σ and consider the hypothesis of testing $H_0 : \theta_1 = \theta_2 = \ldots = \theta_r$. In the classical normal theory, the sample means $\{\overline{X}_i\}$ and variances $\{s_i^2\}$ satisfy

$$\sqrt{n_i}(\overline{X}_i - \theta_i) \quad \sim \quad N(0, \sigma^2), \tag{16}$$

and
$$(n_i - 1)s_i^2 / \sigma_i^2 ~\sim~ \chi_{n_i-1}^2. \tag{17}$$

Then, in the familiar notation $\overline{X}_{..} = \sum n_i \overline{X}_i / \sum n_i$ and $N = \sum n_i$, the well-known variance ratio statistic for testing the homogeneity of r population means is given as,

$$F = \frac{\sum n_i (\overline{X}_i - \overline{X}_{..})^2 / (r-1)}{\sum (n_i - 1)s_i^2 / (N-r)}, \tag{18}$$

which is distributed as $F(r-1, N-r)$ under the null hypothesis and as a noncentral F-distribution under the alternative hypothesis.

Now, let $\tilde{X}_i$ and $\tilde{s}_i^2$, $i = 1, 2, \ldots, r$ be the trimmed means and the Winsorized variances as in (1) and (2) and let

$$\tilde{X}_{..} = \sum w_i^* \tilde{X}_i / \sum w_i^* \quad \text{and} \quad \tilde{s}_p^2 = \left(\sum w_i \tilde{s}_i^2 / V_i^2 \right) / \left(\sum w_i \right), \tag{19}$$

where $w_i^* = n_i / V_i^2$, V_i^2 and w_i are as given in (11) and (12), respectively, be the weighted average of trimmed means and the weighted pooled estimator of the common variance, respectively. Then, a natural robust analog of the classical F-statistic is,

$$\tilde{F}_{pre} = \frac{\sum w_i^* (\tilde{X}_i - \tilde{X}_{..})^2 / (r-1)}{\tilde{s}_p^2}. \tag{20}$$

Note that the $\tilde{F}_{pre}$ statistic reduces to the usual F-statistic when there is zero trimming. Also, note that, In general terms, the estimate of the variance, $\hat{\sigma}^2 = \tilde{s}_p^2$, is the asymptotically minimum variance weighted average of the estimates $\{\hat{\sigma}_i^2 = \tilde{s}_i^2 / V_i^2\}$ obtained from the samples, with weights being inversely proportional to their asymptotic variances of each estimate, and reduces to the usual s_p^2 in case of zero trimming. The asymptotic null distribution of $\tilde{F}_{pre}$ is well approximated by an F-distribution with numerator degrees of freedom $(r-1)$ and denominator degrees of freedom

$$v = 2 \sum w_i. \tag{21}$$

where w_i is given in (12).

The performance, i.e. the type I error control and power properties, of the analog of the classical F-statistic given in (20) is not adequate for moderate samples sizes. Thus, $\tilde{F}_{pre}$ is further modified to be practically useful with sample sizes in practice. This was done by refining (9) and (10) using empirical, Monte Carlo methods and adjusting $\tilde{F}_{pre}$ given in (20), i.e. calibrating its null distribution at the normal distribution as the target family.

Empirical Adjustment of the Null Distribution. For a random sample $X_1, X_2, \ldots, X_n$ from a continuous, symmetric population with distribution function $F[(X - \theta)/\sigma]$, let $\tilde{X}$ be the δ-trimmed mean, a robust estimator of θ. Then for small and moderate size samples it is reasonable to assume that,

$$\sqrt{n}\left(\tilde{X} - \theta\right) \approx N\left(0,\ k(n,\delta)V^2\sigma^2\right), \tag{22}$$

and $\quad \sqrt{n-1}\left(\tilde{s}^2 - m(n,\delta)k(n,\delta)V^2\sigma^2\right) \approx N\left(0, l(n,\delta)U^2\sigma^4\right), \tag{23}$

where the multiplier $k(n,\delta)$, $l(n,\delta)$, and $m(n,\delta)$ are moderate sample size corrections. To estimate these a Monte Carlo experiment with 10,000 replications was undertaken for each combination of samples of size n (from 5 to 50) generated from the target normal family and number of proportion of observations trimmed from each end δ (up to 25% trimming). For each sample, $\tilde{X}$ and $\tilde{s}^2$ were obtained and adjustment $k(n,\delta)$ was obtained by averaging over the Monte Carlo replications as,

$$\hat{k}(n,\delta) = n\hat{Var}(\tilde{X})/[V^2\sigma^2] \tag{24}$$

where $\hat{Var}(\tilde{X})$ is the variance of the $\tilde{X}$'s.

In the same spirit, the moderate size sample size corrections for $l(n,\delta)$ and $m(n,\delta)$ are obtained using 10,000 Monte Carlo simulations as,

$$\hat{l}(n,\delta) = (n-1)\hat{Var}(\tilde{s}^2)/[U^2\sigma^4] \tag{25}$$

and $\quad \hat{m}(n,\delta) = \hat{E}(\tilde{s}^2)/[\hat{k}(n)V^2\sigma^2], \tag{26}$

$\hat{E}(\tilde{s}^2)$ and $\hat{Var}(\tilde{s}^2)$ are the simulated mean and variance of the $\tilde{s}^2$'s and $\hat{k}(n,\delta)$ is given by (24).

The finite sample relations described in equations (24)-(26) are made concrete by using the simplicity dictum of Occam's razor and regressing them on sample size n and trimming proportion δ and obtaining:

$$\hat{k}(n,\delta) = 1.0, \tag{27}$$

$$\hat{m}(n,\delta) = 1.0 - 3.25(\delta/n) + 0.376(\delta/\sqrt{n}) - 32.8(\delta^3/n), \tag{28}$$

$$\hat{l}(n,\delta) = 1.0 + 927(\delta/n^3) - 363(\sqrt{\delta}/n^3) - 1063(\delta^3/n^2). \tag{29}$$

It may be noted that regression equations (27-29) are obtained to ensure that all the three constants tend to 1, as either the sample size $n \to \infty$ or the trimming proportion $\delta \to 0$; $\delta = 0$ corresponds to no trimming and results in the usual F-statistic as in (18) and as $n \to \infty$ no finite sample correction is needed and we get the robust $\tilde{F}_{pre}$-statistic proposed in (20).

Now the robust analog of F given in (20) may be refined in two steps. In the first step the pooled estimate of the common variance σ^2, suppressing n and δ, is obtained as,

$$\tilde{s}_p^2 = \left[\sum w_{i(ref)} \tilde{s}_i^2 / (k_i m_i V_i^2) \right] / \left[\sum w_{i(ref)} \right], \tag{30}$$

a weighted average of the estimated variances, $\hat{\sigma}_i^2 = \tilde{s}_i^2 / (k_i m_i V_i^2)$ with weights $w_{i(ref)} = w_i(k_i^2 m_i^2 / l_i)$, $i = 1, 2, \ldots, r$, inversely proportional to the variances of the $\hat{\sigma}_i^2$.

Then, in the next step, the robust analog of the classical variance ratio F-statistics to be used for moderate size samples is obtained as,

$$\tilde{F} = \frac{\sum w_{i(ref)}^* (\tilde{X}_i - \tilde{X}_{..})^2 / (r-1)}{\tilde{s}_p^2} \tag{31}$$

where $w_{i(ref)}^* = n_i / (k_i V_i^2)$ and $\tilde{X}_{..} = \sum w_{i(ref)}^* \tilde{X}_i / \sum w_{i(ref)}^*$. The null distribution of the test statistic proposed in (31) can be approximated by a $F(r-1, v)$ distribution, where

$$v = 2 \sum w_{i(ref)}. \tag{32}$$

Furthermore, the post-hoc analysis in terms of robust analog of Scheffé's (S-method) simultaneous confidence intervals for all contrasts among the population means can be conducted (Scheffé[26]). Specifically, the simultaneous confidence intervals for all contrasts $\gamma'\theta$, $\gamma \neq 0$ and $\gamma'1 = 0$ given by:

$$\gamma' \tilde{X} - C_\alpha \tilde{s}_p \left(\sum \frac{\gamma_i^2}{n_i} \right)^{\frac{1}{2}} \leq \gamma' \theta \leq \gamma' \tilde{X} - C_\alpha \tilde{s}_p \left(\sum \frac{\gamma_i^2}{n_i} \right)^{\frac{1}{2}}, \quad (33)$$

where $\tilde{X}$ is the vector of r trimmed means and $\tilde{s}_p^2$ is the pooled estimator of variance given in (30), are valid with exact confidence coefficient $(1 - \alpha)100\%$ provided C_α is obtained using the F-distribution approximation for $\tilde{F}$ at (32).

Remark 1: It may be noted that the process of developing robust analog of F test based on trimmed means can be easily modified to incorporate any other robust estimator of θ with an appropriate estimator of the common variance σ^2 as in (30) and obtaining the appropriate fine tuning constants k, l and m as in (24)-(26).

The Trimmed-ANOVA Process. In Summary: Given r samples from (approximately) symmetric populations with means θ_i and common variance σ^2, obtain δ_i-trimmed means and their corresponding Winsorized variances $\tilde{s}_i^2$ using equations (1) and (2), for suitable trimming proportion δ_i. Then, obtain the pooled estimate of the common variance σ^2 given in (30) and the robust $\tilde{F}$-statistic given in (31) and compare it with the null distribution approximation given in (32).

3. Operating Characteristics

The operating characteristics of the trimmed ANOVA $\tilde{F}$-test were empirically studied using an extensive Monte Carlo simulation experiment. The objective was to ascertain its acceptability in terms of validity robustness and efficiency robustness properties. Towards this end 5000 random samples of sizes n in the range of 5-50 from the normal and several heavy tailed populations, under both the null and several non-null assumptions, were simulated and the $\tilde{F}$-test was applied to each samples. Specifically, the populations included: (I). Normal population G, (II). Contaminated normal populations such as $G + 20\% \, 3G$, (III). $G + 20\% \, 4G$, (IV). $G + 10\% \, G/U$, and other long-tailed populations such as (V). t-distribution with 3 $d.f.$ (t_3) and (VI). Cauchy. The notation $G + p\% \, \sigma G$ represents sampling from the unit normal population with probability $(1 - p)$ and from the $N(0, \sigma)$ with probability p. Also, in $G + p\% \, G/U$, also known as the slash distribution, G/U represents the distribution of the ratio of independent standard normal and uniform $U(0,1)$ random variables.

Table 1. Empirical[a] Type I Error Control of the $\tilde{F}$ -Test[b] for $H_0 : \theta_1 = \theta_2 = \theta_3 = 0$.

Sample Sizes			# of Observations Trimmed from each end			Degrees of freedom	Nominal Probabilities (α)		
n_1	n_2	n_3	g_1	g_2	g_3	v	0.01	0.05	0.10
				Normal Population					
20	20	20	0	0	0	57.00	0.0100	0.0486	0.1010
20	20	20	3	3	3	32.06	0.0114	0.0526	0.0990
20	20	20	5	5	5	18.40	0.0078	0.0458	0.0932
50	50	50	0	0	0	147.00	0.0092	0.0514	0.1028
50	50	50	7	7	7	89.12	0.0100	0.0502	0.1018
50	50	50	10	10	10	68.55	0.0080	0.0416	0.0984
10	20	50	0	0	0	77.00	0.0116	0.0534	0.1052
10	20	50	1	2	5	54.09	0.0132	0.0524	0.1018
10	20	50	2	5	10	32.34	0.0072	0.0434	0.0918
				$G + 20\%\ 4G$					
20	20	20	0	0	0	57.00	0.0076	0.0476	0.1032
20	20	20	3	3	3	32.06	0.0068	0.0404	0.0872
20	20	20	5	5	5	18.40	0.0082	0.0426	0.0876
50	50	50	0	0	0	147.00	0.0072	0.0440	0.0982
50	50	50	7	7	7	89.12	0.0106	0.0488	0.1006
50	50	50	10	10	10	68.55	0.0070	0.0422	0.0910
10	20	50	0	0	0	77.00	0.0116	0.0558	0.0992
10	20	50	1	2	5	54.09	0.0148	0.0580	0.1142
10	20	50	2	5	10	32.34	0.0106	0.0494	0.1006
				$G + 10\%\ G/U$					
20	20	20	0	0	0	57.00	0.0036	0.0270	0.0722
20	20	20	3	3	3	32.06	0.0084	0.0432	0.0908
20	20	20	5	5	5	18.40	0.0086	0.0466	0.0948
50	50	50	0	0	0	147.00	0.0020	0.0238	0.0620
50	50	50	7	7	7	89.12	0.0094	0.0480	0.0996
50	50	50	10	10	10	68.55	0.0092	0.0472	0.0962
10	20	50	0	0	0	77.00	0.0096	0.0584	0.1096
10	20	50	1	2	5	54.09	0.0090	0.0448	0.0962
10	20	50	2	5	10	32.34	0.0092	0.0472	0.0936

[a] Based on Monte Carlo experiment with 5000 replications.
[b] Test Described in Section 2.1 and summarized at the end of Section 2 with Null distribution approximation at (32).

Validity Robustness. It is evaluated in terms of type I error control across the relevant populations. Obviously, for the test based on an approximation to the null distribution cannot have exact type I error with moderate size samples, but is expected to be close to the nominal value in case of both the normal and moderately heavy tailed populations. In view of the space limitation, a small selection of the results appears in Table 1 whereas the performance of the test statistic over a broader range of populations is provided in figures which can be accessed from the link http://www.stjude.org/trimmedanova-nullfigures.

- From Table 1 and figures as mentioned in the above link, it is seen that when the normality assumption holds the type I error control with the $\tilde{F}$-test at (31) with or without trimming are comparable, except in case of small samples and heavy trimming when it becomes conservative.

- Similarly, from Table 1 and the link mentioned above, an improvement in the type I error control with the $\tilde{F}$-test is seen when the populations are contaminated normal type. Furthermore, the improvement is enhanced with increasing levels of contamination together with higher levels of trimming. Similar conclusions can be drawn for the heavy tailed, Student's t-distribution with 3 $d.f.$, population.

- In a similar manner, From Table 1 and the link mentioned above, a dramatically superior type I error control is seen with the $\tilde{F}$-test as compared to the F-test when the populations are extremely heavy tailed population such as Cauchy or Slash, i.e. contamination with G/U. For example, when the underlying distributions are contaminated with G/U and the samples sizes are 50 for the three groups then the estimate of the Monte Carlo estimate of the 5% level of significance for the usual F statistic is 0.0238 whereas with the robust $\tilde{F}$ with 20% trimming the estimate is 0.0472.

Efficiency Robustness: As Scheffé[26] notes, a robust test is expected to lose some power when the population is normal, but should provide substantial gain at most non-normal populations. The Monte Carlo experiment described above, in this sense, confirms robustness of $\tilde{F}$-test. A brief selection of the power function of the $\tilde{F}$-test is given in Table 2 and Figures 1 and 2; a more detailed evaluation over broader range of populations is provided in the power figures which can be accessed from the link http://www.stjude.org/trimmedanova-powerfigures. Some of the observations gleaned from the results of Monte Carlo study are now briefly summarized:

Table 2. Empirical[a] Power Function of the $\tilde{F}$ -Test[b] for $H_1 : \Delta = a \Rightarrow \theta_1 = 0, \theta_2 = \theta_3 = a$.

Sample Sizes			# of Observations Trimmed from each end			Degrees of freedom	$\alpha = 0.05$		$\alpha = 0.01$	
n_1	n_2	n_3	g_1	g_2	g_3	ν	$\Delta = 0.5$	$\Delta = 1.0$	$\Delta = 0.5$	$\Delta = 1.0$
						Normal Population				
20	20	20	0	0	0	57.00	0.324	0.905	0.135	0.741
20	20	20	3	3	3	32.06	0.292	0.855	0.111	0.649
20	20	20	5	5	5	18.40	0.259	0.787	0.095	0.521
50	50	50	0	0	0	147.00	0.738	1.000	0.511	0.996
50	50	50	7	7	7	89.12	0.680	0.999	0.444	0.993
50	50	50	10	10	10	68.55	0.662	0.999	0.408	0.990
10	20	50	0	0	0	77.00	0.233	0.739	0.087	0.498
10	20	50	1	2	5	54.09	0.217	0.721	0.078	0.462
10	20	50	2	5	10	32.34	0.205	0.648	0.068	0.392
						$G + 20\%\ 4G$				
20	20	20	0	0	0	57.00	0.117	0.370	0.034	0.169
20	20	20	3	3	3	32.06	0.187	0.634	0.060	0.361
20	20	20	5	5	5	18.40	0.177	0.601	0.049	0.325
50	50	50	0	0	0	147.00	0.256	0.729	0.092	0.516
50	50	50	7	7	7	89.12	0.464	0.971	0.236	0.904
50	50	50	10	10	10	68.55	0.466	0.973	0.241	0.900
10	20	50	0	0	0	77.00	0.098	0.256	0.026	0.098
10	20	50	1	2	5	54.09	0.136	0.456	0.042	0.230
10	20	50	2	5	10	32.34	0.145	0.474	0.044	0.235
						$G + 10\%\ G/U$				
20	20	20	0	0	0	57.00	0.170	0.507	0.061	0.329
20	20	20	3	3	3	32.06	0.260	0.787	0.094	0.569
20	20	20	5	5	5	18.40	0.211	0.724	0.073	0.459
50	50	50	0	0	0	147.00	0.266	0.595	0.129	0.478
50	50	50	7	7	7	89.12	0.607	0.997	0.366	0.985
50	50	50	10	10	10	68.55	0.590	0.995	0.342	0.977
10	20	50	0	0	0	77.00	0.130	0.355	0.040	0.186
10	20	50	1	2	5	54.09	0.192	0.636	0.069	0.380
10	20	50	2	5	10	32.34	0.183	0.600	0.059	0.336

[a] Based on Monte Carlo experiment with 5000 replications.
[b] Test described in Section 2.1 and summarized at the end of Section 2 with Null Distribution approximation at (32).

164

Three Samples from Normal Population[a]

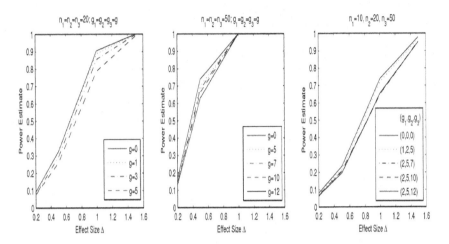

Three Samples from $G + 20\%\, 4G$ Population[a].

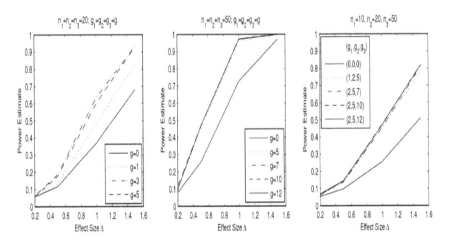

Fig. 1. Empirical[b] Power Function of the $\tilde{F}$ -Test of Section 2.1 for three samples at $\alpha = 0.05$. Effect size $\Delta = a \Rightarrow \theta_1 = 0, \theta_2 = \theta_3 = a$.

---

[a] For details see Section 3.
[b] Based upon a Monte Carlo experiment with 5000 replications.

Three Samples from $G + 10\%\, G\,/\,U$ Population[a].

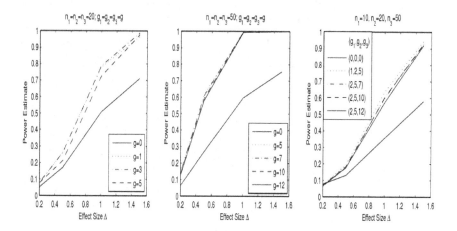

Three Samples from Cauchy Distribution[a].

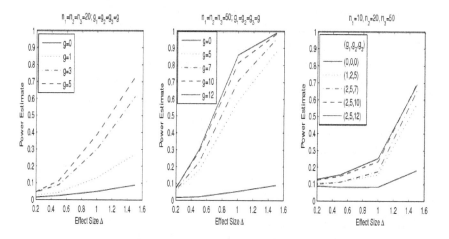

Fig. 2. Empirical[b] Power Function of the $\tilde{F}$ -Test of Section 2.1 for three samples at α =0.05.
Effect size $\Delta = a \Rightarrow \theta_1 = 0,\, \theta_2 = \theta_3 = a.$

---

[a] For details see Section 3.
[b] Based upon a Monte Carlo experiment with 5000 replications.

- Figure 1 and Table 2, as expected, show minor power loss with the $\tilde{F}$-test compared to the F-test. For example, when the sample sizes are $n_1 = n_2 = n_3 = 50$ and $\Delta = 0.5$, $(\Delta = a \Rightarrow \theta_1 = 0, \theta_2 = \theta_3 = a)$ the power for the regular F-test $(g_1 = g_2 = g_3 = 0)$ at 5% level of significance is about 74% whereas for the $\tilde{F}$-test with roughly 15% $(g_1 = g_2 = g_3 = 7)$ and 20% $(g_1 = g_2 = g_3 = 10)$ trimming the power estimates are 68% and 66%, respectively.

- In case of contaminated Normal populations, Figure 1, Table 2 and the link mentioned above, a clear power gain of the $\tilde{F}$-test over the F-test is seen. For example when the population is $G + 20\%$ $4G$, sample sizes are $n_1 = n_2 = n_3 = 50$ and $\Delta = 1.0$ then the power estimates with no trimming and with 14% and 20% trimming are 73%, 97% and 97%, respectively.

- Similarly, from Figure 2, Table 2 and the link mentioned above, a significant power advantage is seen for the very heavy tailed populations, such as Student's t-distribution with 3 $d.f.$, Slash and Cauchy distributions. In these cases, the power superiority of the $\tilde{F}$-test over the F-test increases as the trimming proportion increases. This phenomenon is especially clear in case of Cauchy or Slash populations. For example when a Standard normal distribution is 10% contaminated with G/U, $n_1 = n_2 = n_3 = 50$ and $\Delta = 1.0$ then the power estimates with no trimming and with 14% and 20% trimming are 60%, 99% and 99%, respectively.

4. Conclusions

In this paper we have presented the $\tilde{F}$-test, based on trimmed means, as a robust alternative to the classical ANOVA test. An extensive Monte Carlo study shows that it is validity robust, i.e. its null distribution approximation provides reasonable type I error control when the normality assumption is satisfied as well as when the populations are heavy tailed. Furthermore, the same simulation study shows that it is efficiency robust, i.e. at the cost of small power loss when the normality assumption holds; it provides substantial power advantage when the populations are heavy tailed. More importantly, the $\tilde{F}$-test can be used exactly as the normal theory F-test for the post-hoc inference, i.e. for construction of Scheffé's (S-method) simultaneous confidence intervals on all contrasts among the population means. As a general recommendation for the practice, based on extensive Monte Carlo study, we suggest that the $\tilde{F}$-test based on 15% trimming may be appropriate and should be used.

Acknowledgments

The research work of Deo Kumar Srivastava was in part supported by the American Lebanese Syrian Associated Charities. The authors are thankful to the referees for their constructive comments and suggestions. The authors are also thankful to Ms. Peggy Vandiveer for formatting the manuscript.

References

1. D. F. Andrews, P. J. Bickel, F. R. Hampel, P. J. Huber, W. H. Rogers, and J. W. Tukey. *Robust Estimators of Location, Survey and Advances* (Princeton University Press, Princeton, NJ, 1972).
2. B. Bertaccini, and R. Varriale. *Computational Statistics & Data Analysis* **51**, 5172 (2007).
3. G. E. P. Box and D. R. Cox. *Journal of the Royal Statistical Society* **B26**, 211 (1964).
4. J. V. Bradley. *British Journal of Mathematical and Statistical Psychology* **31**, 144 (1978).
5. M. B. Brown, and A. B. Forsythe. *Technometrics* **16**, 129 (1974).
6. D. Draper. *Statistical Science*, **3**, 239 (1988).
7. C. W. Dunnett. *Communications in Statistics – Theory and Methods* **11**, 2611 (1982).
8. K. Y. Fung, and H. Tam. *The Statistician* **37**, 387 (1988).
9. A. Garcia-Perez. *Test* **17**, 350 (2008).
10. A. Gelman. *The Annals of Statistics* **33**, 1 (2005).
11. A. Giloni, S. Seshadri, and J. Simonoff. *International Journal of Productivity and Quality Management* **1**, 306 (2006).
12. F. R. Hampel, E. M. Ronchetti, P. J. Rousseeuw and W. A. Sathel. *Robust Statistics* (John Wiley & Sons, New York, 1986).
13. P. L. Hsu. *Statistical Research Memoirs* **2**, 1 (1938).
14. P. J Huber. *Nonparametric Techniques in Statistical Inference*, Ed.: M. L. Puri (Cambridge University Press, Cambridge, England, 1970), 453.
15. P. J. Huber. *Robust Statistics* (John Wiley & Sons, New York, 1981).
16. G. S. James. *Biometrika* **38**, 324 (1951).
17. W. H. Kruskal and W. A. Wallis. *Journal of the American Statistical Association* **57**, 583 (1952).
18. W. H. Lawton. *Annals of Institute of Mathematical Statistics* **36**, 1521 (1965).
19. H. Lee, and K. Y. Fung. *Sankhyā* **B47**, 186 (1985).
20. W-M Luh, and J-H Guo. *British Journal of Mathematical and Statistical Psychology* **52**, 303 (1999).

168

21. F. Mosteller, and J. W. Tukey. *Data Analysis and Regression* (Addison-Wesley, Reading, MA, 1977).
22. A. Mudholkar, G. S. Mudholkar, and D. K. Srivastava. *Communications in Statistics – Theory and Methods* **20**, 1345 (1991).
23. G. S. Mudholkar, and K. R. Patel. *Journal of the Indian Statistical Association* **17**, 197 (1979).
24. K. R. Patel, G. S. Mudholkar, and J. L. I. Fernando. *Journal of the American Statistical Association* **83**, 1203 (1988).
25. P. Prescott. *Journal of the American Statistical Association* **73**, 133 (1978).
26. H. Scheffé. *The Analysis of Variance* (John Wiley & Sons, New York, 1959).
27. R. M. Schrader and T. P. Hettmansperger. *Biometrika* **67**, 93 (1980).
28. R. M. Schrader and J. W. McKean. *Communications in Statistics – Theory and Methods* **6**, 879 (1977).
29. G. R. Shorack. *The Annals of Statistics* **2**, 661 (1974).
30. G. R. Shorack. *Statistica Neerlandica* **30**, 119 (1976).
31. D. K. Srivastava, G. S. Mudholkar, and A. Mudholkar. *Journal of Applied Statistics* **19**, 405 (1992).
32. S. M. Stigler. *The Annals of Statistics* **1**, 472 (1973).
33. W. Y. Tan and M. A. Tabatabai. *Communications in Statistics – Simulation and Computation* **14**, 1007 (1985).
34. W. Y. Tan and M. A. Tabatabai. *Communications in Statistics – Simulation and Computation* **15**, 733 (1986).
35. D. Thorburn. *Biometrika* **63**, 305 (1976).
36. J. W. Tukey. A survey of sampling from contaminated distributions. *Contributions to Probability and Statistics – Essays in Honor of Harold Hotelling* (Stanford University Press, CA, 1960).
37. J. W. Tukey. *The Collected Works of John W. Tukey*, Ed.: L. V. Jones (Wadsworth & Brooks/Cole, Monterey, CA, 1962), 391.
38. J. W. Tukey and D. H. McLaughlin. *Sankhyā: The Indian Journal of Statistics* **A25**, 331 (1963).
39. B. L. Welch. *Biometrika* **38**, 330 (1951).
40. R. R. Wilcox, H. J. Keselman, and R. K. Kowalchuk. *British Journal of Mathematical and Statistical Psychology* **51**, 123 (1998).
41. K. K. Yuen. *Biometrika* **61**, 165 (1974).
42. K. K. Yuen and W. J. Dixon. *Biometrika* **60**, 369 (1973).

BAYESIAN PREDICTIVE INFERENCE UNDER
SEQUENTIAL SAMPLING WITH SELECTION BIAS

B. NANDRAM

Department of Mathematical Sciences, Worcester Polytechnic Institute,
100 Institute Road, Worcester, MA 01609, England
E-mail: balnan@wpi.edu

D. KIM

Department of Statistics, Kyungpook National University,
1370 Sangpeok-dong, Buk-gu, Daegu 702-701, Korea
E-mail: dalkim@knu.ac.kr

A finite population, which consists of clusters, is sampled sequentially, and Bayesian predictive inference is used to infer about the number of all members and the proportion of members with a certain characteristic in the entire finite population. However, the sequential sample is biased; this is a nonstandard selection bias problem because all selection probabilities are unknown. We construct a hierarchical Bayesian model, a nonignorable selection model, to do Bayesian predictive inference. The nonignorable selection model has a mechanism in which the selection probabilities of the clusters are proportional to the cluster sizes. We discuss probabilistic features of the nonignorable selection model, and show that statistical inference is credible. We use Monte Carlo methods (not Markov chain) to get samples (used for posterior inference) from the nonignorable selection model. We use an example on cystic fibrosis to illustrate Bayesian predictive inference. For sensitivity analysis, we compare the nonignorable selection model with an ignorable selection model and a second nonignorable selection model that has the unknown selection probabilities proportional to the number of affected siblings.

1. Introduction

We study a sequential sampling plan for finite population sampling with a selection bias mechanism. Clusters are sampled sequentially and a small sample (one or two) of the units within the cluster is taken. Then, these units are tested. If all tested units are negative, no further testing is done in this cluster and the information for this cluster is discarded (i.e., no record is kept). If there is at least one positive, all units in the sample and the rest of the cluster (if any) are also tested, and all information for this cluster is recorded. The next cluster is then sampled, and the procedure continues until all clusters are sampled in this sequential sampling plan. It is of interest to predict the total number of positive units and the proportion of positive units in the population. Both of these predictions can be badly biased because if the initial sampled units are all negatives, the entire cluster is discarded and it is possible that this cluster may contain positive units as well.

A sample of n clusters is selected sequentially from the finite population as follows. When the first positive cluster is recorded, we assume that there are $N_1 - 1$ clusters that have been sampled with negative results. That is, the first positive unit appears in the N_1^{th} cluster. Continue sampling and start counting again, the procedure continues and the number of clusters sampled until the next cluster shows a positive unit is N_2. This continues until the whole finite population is exhausted. Let n denote the random number of clusters sampled. Thus, the complication here is that n and the N_i are random variables, and the number of positive units in the discarded clusters are unknown. However, it is a common practice to make inference conditional on n, and we continue to do so. There are many applications of this framework.

Although we wish to construct a general purpose methodology, to make our discussion more concrete, we describe an example in population genetics. Here the word, family, is more appropriate than cluster. When a disease is rare, it is inefficient to draw a random sample of families, and only families with at least one affected sibling enter a study. Families who come to the attention of an investigator are said to be ascertained. Thus, there is an ascertainment bias and corrections are, therefore, needed in the study. One can view ascertainment as a sampling process in which unascertained families are discarded one by one until a family is finally ascertained. Families are ascertained through their affected siblings; see Lange.[10] If all siblings who first appear at a medical unit are not affected, their information is discarded and no more testing is done in that family; otherwise the entire family is tested. Ascertainment continues until another family is found with at least one affected sibling, and the process continues until the population is exhausted. After the $(i - 1)^{th}$ affected family is found, the number of families tested until the ith affected family, N_i, is observed, is clearly random variable. Also no record of the unascertained families comes to the attention of the investigator. The number of ascertained families n becomes known only after the entire population has been sampled; so that

n is a random variable. Moreover, the sizes and the number of the families affected are unobservables; these are observed for the ascertained families. In a Bayesian analysis all unobservables are random variables. The total number affected in the entire population and their proportion are of interest. We have an illustrative example in population genetics on cystic fibrosis which we discuss later.

Let a_{ij}, $j = 1, \ldots, N_i$, $i = 1, \ldots, n$, denote the number of positive units initially sampled from the jth cluster. (These positive units are called *probands* in medical studies.) Here $a_{ij} = 0, j = 1, \ldots, (N_i - 1)$, and $a_{iN_i} \geq 1$, $i = 1, \ldots, n$. It is convenient to take $a_{iN_i} \equiv a_i$. Here the cluster size s_i are observed but the cluster sizes s_{ij}, $j = 1, \ldots, (N_i - 1)$ are not observed. Similarly, the numbers of positive responses r_i are observed but r_{ij}, $j = 1, \ldots, (N_i - 1)$ are not observed. Thus, the N_i, (r_{ij}, s_{ij}), $j = 1, \ldots, (N_i - 1)$, are unknown. In this article, the key parameters of interest are

$$R = \sum_{i=1}^{n} r_i + \sum_{i=1}^{n} \sum_{j=1}^{N_i-1} r_{ij} \text{ and } P = \frac{\sum_{i=1}^{n} r_i + \sum_{i=1}^{n} \sum_{j=1}^{N_i-1} r_{ij}}{\sum_{i=1}^{n} s_i + \sum_{i=1}^{n} \sum_{j=1}^{N_i-1} s_{ij}}. \quad (1)$$

We use Bayesian predictive inference to predict R and P in (1). This can be obtained using models to predict the r_{ij} and s_{ij}, $j = 1, \ldots, (N_i - 1)$, $i = 1, \ldots, n$.

There are several articles on selection bias for finite population. However, the one that is most pertinent here is the article by Chambers, Dorfman and Wang[2]; see the references therein. Chambers, Dorfman and Wang[2] assume that one wishes to model the population process that yields the finite population of survey variables. They assume that the only information about the survey design available to the survey analyst is the set of first-order inclusion probabilities for the sampled units. They focus on and generalize an example presented by Krieger and Pfeffermann[9] whose objective is to investigate inferential methods when there is a selection bias. We note that the model of Chambers, Dorfman and Wang[2] is of limited scope in our problem because it assumes that the finite population is generated as a random sample from a normal distribution. It is clear that normality cannot be assumed in our problem but the framework of Chambers, Dorfman and Wang[2] is important. Under probability proportional to size (PPS) sampling, Nandram *et al.*[17] used the model of Chambers, Dorfman and Wang[2] to do Bayesian predictive inference when a transformation of a continuous response is needed. Also for PPS sampling with a binary response, Nandram[12] implemented surrogate sampling techniques to provide simulated random samples by using a model which reverses the selection bias. Finally we note that Nandram and Choi,[15] generalizing a hierarchical Bayesian model of Malec, Davis and Cao,[11] include both selection bias and nonresponse bias into their model.

We use two hierarchical Bayesian models to predict these parameters. Generally, a selection model is called nonignorable if the response and the selection indicators are correlated; otherwise the model is called ignorable. The ignorable selection model does not recognize the fact that $a_{ij} = 0, j = 1, \ldots, N_i - 1$. This is incorporated into the nonignorable selection model via a Bernoulli sampling scheme. That is, the nonignorable selection model has a piece of information which is not included in the ignorable selection model. Thus, the two models are expected to give different answers, and we believe that the answers from the nonignorable selection model should be preferred. In the Bernoulli sampling scheme we assume that the selection probabilities are proportional to either the r_{ij} or the s_{ij}. Thus, another complication is that both of these quantities are unknown.

This article has four additional sections. In Section 2 we describe the Bayesian methodology. Specifically, we describe the two Bayesian models, the ignorable selection model and the nonignorable selection model. In Section 3 we show how to fit the two models using Monte Carlo methods. An important feature of our methodology is that we do not need to use Markov chain Monte Carlo methods. In Section 4 we describe an illustrative example on cystic fibrosis, a genetic application. However, our framework has greater generality. Section 5 has concluding remarks.

2. Ignorable and Nonignorable Selection Models

The ignorable selection model does not take the selection bias into account while the nonignorable selection model does.

First, we describe the ignorable selection model because it is used to construct the nonignorable selection model. We assume

$$a_{ij} \mid r_{ij}, \pi, N_i \overset{ind}{\sim} \text{Binomial}(r_{ij}, \pi), \tag{2}$$

and

$$r_{ij} \mid s_{ij}, p, N_i \overset{ind}{\sim} \text{Binomial}(s_{ij}, p), \tag{3}$$

$j = 1, \ldots, N_i, i = 1, \ldots, n$. Note that $N = \sum_{i=1}^{n} N_i$ is assumed to be unknown. In genetics π is called the *proband probability* and p is called the *segregation probability*. It is sensible to use a Poisson distribution to model the s_{ij} but because the s_{ij} are positive we need to truncate the

Poisson distribution. We say that a nonnegative discrete random variable $T \sim TP(\lambda)$ if $P(T = t \mid \lambda) = \lambda^t e^{-\lambda}/t!(1 - e^{-\lambda})$, $t \geq 1$. Thus, we assume

$$s_{ij} \mid \lambda, N_i \overset{ind}{\sim} TP(\lambda), s_{ij} \geq 1, \; j = 1, \ldots, N_i. \tag{4}$$

It is worth noting that the conditional distributions in (2), (3) and (4) depend on the N_i through the index j. Because there are many N_i we prefer not to use a noninformative prior (Raftery[19]) for each of the N_i. In this case it is more sensible to assume that the N_i come from a common distribution. However, noting that in (2) $P(a_{ij} > 0 \mid \pi, p, \lambda) = (1 - e^{-\pi p \lambda}/(1 - e^{-\lambda})$, we have used the geometric distribution with success probability $(1 - e^{-\pi p \lambda}/(1 - e^{-\lambda})$. That is,

$$N_i \mid \pi, p, \lambda, \overset{iid}{\sim} \text{Geometric}\{\frac{1 - e^{-\pi p \lambda}}{1 - e^{-\lambda}}\}, N_i \geq 1, \; i = 1, \ldots, n. \tag{5}$$

Finally, a priori we assume that π, p and λ are independent with

$$\pi(\pi, p, \lambda) \propto \frac{1}{\{\pi(1 - \pi)p(1 - p)\lambda\}^k}, 0 < \pi, p < 1, \lambda > 0, \tag{6}$$

where k is to be chosen. Note that $k = .5$ gives us Jeffreys' prior, a standard noninformative prior. Other values of k are also sensible (e.g., $k = 1$ gives improper priors for π and p and $k = 0$ gives uniform priors for π and p; for any value of k the prior for λ is improper).

Next, our nonignorable selection model is described. A reasonable assumption is that the selection probabilities are proportional to r_{ij}, $i = 1, \ldots, n$, $j = 1, \ldots, N_i$. Letting $\bar{r} = \sum_{i=1}^{n} \sum_{j=1}^{N_i} r_{ij} / \sum_{i=1}^{n} N_i$ and $\bar{N} = \sum_{i=1}^{n} N_i/n$, under sampling without replacement, the families are selected with probabilities $\frac{r_{ij}}{\bar{N}\bar{r}}(= \frac{r_{ij}}{\bar{r}\bar{N}})$ which follows from a standard internal consistency condition used by survey samplers (Cochran[3]). This is true because the selection probabilities are proportional to the r_{ij} and under sampling without replacement they must sum to n. Let $I_{ij} = 1$ for selection, and $I_{ij} = 0$ for no selection. Thus,

$$P(\underset{\sim}{I} \mid \underset{\sim}{r}, \underset{\sim}{N}) = \prod_{i=1}^{n} \left\{ \left(\prod_{j=1}^{N_i - 1} \left(1 - \frac{r_{ij}}{\bar{N}\bar{r}}\right) \frac{r_i}{\bar{N}\bar{r}} \right) \right\}. \tag{7}$$

The probabilistic structure in (7) for the selection process is similar to the one specified by Chambers, Dorfman and Wang,[2] but in (7) only the sampled r_i are observed.

Adding (7) into (2), (3), (4), (5) of the ignorable selection model, we see the joint distribution of $I, a, r, s, N \mid \pi, p, \lambda$ is

$$
P(\underset{\sim}{I}, \underset{\sim}{a}, \underset{\sim}{r}, \underset{\sim}{s}, \underset{\sim}{N} \mid \pi, p, \lambda) = \prod_{i=1}^{n} \prod_{j=1}^{N_i-1} \left\{ (1 - \frac{r_{ij}}{\bar{N}\bar{r}}) \frac{\lambda^{s_{ij}} e^{-\lambda}}{s_{ij}!(1 - e^{-\lambda})} \right.
$$

$$
\times \left. \binom{r_{ij}}{a_{ij}} \pi^{a_{ij}} (1 - \pi)^{r_{ij} - a_{ij}} \binom{s_{ij}}{r_{ij}} p^{r_{ij}} (1 - p)^{s_{ij} - r_{ij}} I(a_{ij} = 0, I_{ij} = 0) \right\}
$$

$$
\times \prod_{i=1}^{n} \left\{ [\frac{e^{-\pi p \lambda} - e^{-\lambda}}{1 - e^{-\lambda}}]^{N_i - 1} [\frac{1 - e^{-\lambda \pi p}}{1 - e^{-\lambda}}][\frac{r_i}{\bar{N}\bar{r}} \frac{\lambda^{s_i} e^{-\lambda}}{s_i!(1 - e^{-\lambda})}] \right.
$$

$$
\times \left. \binom{r_i}{a_i} \pi^{a_i} (1 - \pi)^{r_i - a_i} \binom{s_i}{r_i} p^{r_i} (1 - p)^{s_i - r_i} \right\}. \tag{8}
$$

Using Bayes' theorem, it follows from the prior (6) and the likelihood (8) that the joint posterior density (excluding the normalization constant) of $r^*, s^*, N, \pi, p, \lambda \mid I, a, r_o, s_o$, where r^* and s^* are unobserved values of r_{ij} and s_{ij} and r_o and s_o are observed values of the r_i and s_i, is easy to write down. Note that $\underset{\sim}{I}$ and a are observed for all families and a makes s redundant (i.e., if a is known, I becomes known but not vice versa).

Thus, letting $s = \sum_{i=1}^{n} s_i$, $r = \sum_{i=1}^{n} r_i$, $a = \sum_{i=1}^{n} a_i$, the joint posterior density is

$$
\pi(\underset{\sim}{r}^*, \underset{\sim}{s}^*, \underset{\sim}{N}, \pi, p, \lambda \mid \underset{\sim}{I}, \underset{\sim}{a}, \underset{\sim}{r}_o, \underset{\sim}{s}_o) \propto
$$

$$
T(\underset{\sim}{r}, \underset{\sim}{N}, \lambda, \pi, p) \pi_a^{(ns)}(\underset{\sim}{r}^*, \underset{\sim}{s}^*, \underset{\sim}{N} \mid \pi, p, \lambda, r_o, s_j, a = 0) \pi_a^{(s)}(\pi, p, \lambda \mid \underset{\sim}{r}_o, \underset{\sim}{s}_o, a), \tag{9}
$$

where

$$
T(\underset{\sim}{r}, \underset{\sim}{N}, \lambda, \pi, p) = \left(\frac{1 - e^{-\lambda p}}{1 - e^{-\lambda}} \right)^n \prod_{i=1}^{n} \left\{ \left[\prod_{j=1}^{N_i-1} \left(1 - \frac{r_{ij}}{\bar{N}\bar{r}} \right) \right] \frac{r_i}{\bar{N}\bar{r}} \right\},
$$

$$
\pi_a^{(s)}(\pi, p, \lambda \mid \underset{\sim}{r}_0, \underset{\sim}{s}_0, a) \propto \frac{\lambda^{s-1} e^{-n\lambda} \pi^{a-1} (1 - \pi)^{r-a-1} p^{r-1} (1 - p)^{s-r-1}}{[1 - e^{-\lambda \pi p}]^n},
$$

and

$$
\pi_a^{(ns)}(\underset{\sim}{r}^*, \underset{\sim}{s}^*, \underset{\sim}{N} \mid \pi, p, \lambda, r_0, s_0, a = 0) \propto \prod_{i=1}^{n} \left[\frac{e^{-\pi p \lambda} - e^{-\lambda}}{1 - e^{-\lambda}} \right]^{N_i - 1}
$$

$$\times \prod_{i=1}^{n} \prod_{j=1}^{N_i-1} \frac{\lambda^{s_{ij}} e^{-\lambda}}{s_{ij}!(1 - e^{-\lambda})} \binom{s_{ij}}{r_{ij}} [(1 - \pi)p]^{r_{ij}} (1 - p)^{s_{ij} - r_{ij}}.$$

For $N_i = 1$ we set the corresponding terms equal 1.

It is interesting to look at the form of the joint posterior density in (9). First, it is the exact form (i.e., there are no approximations in it). Second, it is the product of three terms, $T(r, N, \lambda, \pi, p)$, $\pi_a^{(s)}(\pi, p, \lambda \mid r_0, s_0, a)$ and $\pi_a^{(rs)}(r^*, s^*, N \mid \pi, p, \lambda, r_0, s_0, a = 0)$. In this form, the product of $\pi_a^{(s)}(\pi, p, \lambda \mid r_0, s_0, a)$ and $\pi_a^{(\tilde{r}s)}(\tilde{r}^*, s^*, N \mid \pi, p, \lambda, r_0, s_0, a = 0)$ is an approximation to the joint posterior density and $T(r, N, \lambda, \pi, p)$ is an adjustment to this approximation. Note specifically that we have kept the term, $[1 - e^{-\lambda \pi p}]^n$ in $\pi_a^{(s)}(\pi, p, \lambda \mid r_0, s_0, a)$, to help adjusting for the selection bias. Thus, one might believe that $\pi_a^{(\tilde{s})}(\pi, p, \lambda \mid r_0, s_0, a)$ and $\pi_a^{(rs)}(r^*, s^*, N \mid \pi, p, \lambda, r_0, s_0, a = 0)$ are, respectively, approximations to the true distributions, $\pi^{(s)}(\pi, p, \lambda \mid r_0, s_0, a)$ and $\pi^{(rs)}(r^*, s^*, N \mid \pi, p, \lambda, r_0, s_0, a = 0)$. Our computations show that the overall approximation is reasonable.

3. Bayesian Computations

It is natural to use a reversible jump sampler (e.g., Green[8]) to do the computation because the numbers of r_{ij} and s_{ij} change with N_i. However, we do not need Markov chain Monte Carlo methods (Robert and Casella[20]). We can obtain a much simpler algorithm using Monte Carlo methods with random draws (not a Markov chain). Our procedure is described in the following three steps:

(a) Draw π, p, λ from $\pi_a^{(s)}(\pi, p, \lambda \mid r_0, s_0, a)$;
(b) Draw r^*, s^*, N from $\pi_a^{(ns)}(r^*, s^*, N \mid \pi, p, \lambda, r_0, s_0, a = 0)$;
(c) Use the sampling importance resampling (SIR) algorithm to subsample the samples drawn in (a) and (b).

Note that (a) is done using only the observed data, so this part does not constitute a varying dimension problem. Also, although (b) has varying dimension, the simple composition method is used to accomplish this task. Thus, a reversible jump sampler is not needed. In (c) the SIR algorithm helps us to avoid the more sophisticated Markov chain Monte Carlo methods; see Gelman *et al.*[6] (pg. 316).

We draw a large sample $\tilde{M} \approx 10,000$ in (a) and (b), denoted by $(\pi^{(h)}, p^{(h)}, \lambda^{(h)}, r_{ij}^{(h)}, s_{ij}^{(h)}, i = 1, \ldots, n, j = 1, \ldots, N_i^{(h)})$, $h = 1, \ldots, \tilde{M}$. Then, in (c) we draw (subsampling) a sample of size M without replacement from the initial sample of size $\tilde{M}$ with weights

$$w^{(h)} \propto \left[\frac{1 - e^{-\lambda^{(h)}\pi^{(h)}p^{(h)}}}{1 - e^{-\lambda^{(h)}}} \right]^n \prod_{i=1}^{n} \left\{ \frac{r_i^{(h)}}{\bar{N}^{(h)}\bar{r}^{(h)}} \prod_{j=0}^{N_i^{(h)}} (1 - \frac{r_{ij}^{(h)}}{\bar{N}^{(h)}\bar{r}^{(h)}}) \right\},$$

$h = 1, \ldots, \tilde{M}$, where the $w^{(h)}$ are not functions of $s_{ij}^{(h)}$. The three steps (a), (b) and (c) use Monte Carlo methods without Markov chain, thereby avoiding the reversible jump algorithm (e.g., Green[8]), a more complex algorithm. Note that we need $\{(1 - e^{-\lambda\pi p})/(1 - e^{-\lambda})\}^n$ to be bounded for proper performance of the SIR algorithm. This is true because $(1 - e^{-\lambda\pi p})/(1 - e^{-\lambda})$ is a probability.

Note that fitting of the ignorable selection model can proceed in the exactly same manner except that

$$w^{(h)} \propto \left(\frac{1 - e^{-\lambda^{(h)}\pi^{(h)}p^{(h)}}}{1 - e^{\lambda^{(h)}}} \right)^n, \ h = 1, \ldots, \tilde{M}.$$

We have used a 10% subsampling (i.e., we subsampled $M = 1,000$ of the $\tilde{M} = 10,000$ draws.) This is a reasonable subsampling rate; see Smith and Gelfand[21]. In our illustrative example, we have also used a 1% subsampling rate, and we have seen very small changes. However, this takes a lot more time and is not necessary because we are using random samples, not Markov chain, so these values of M and $\tilde{M}$ are also the effective sample sizes. Our algorithm is optimal at 10% subsampling. Of course, in other applications one would need to experiment with the subsampling rate.

The key question is how to do (a) and (b). Let us describe (b) first. In (b) it can be shown that

$$2N_i - 1 \mid \pi, p, \lambda \overset{ind}{\sim} \text{Geometric} \left\{ \frac{1 - e^{-\lambda\pi p}}{1 - e^{-\lambda}} \right\}, \tag{10}$$

$$s_{ij} \mid N_i, \pi, p, \lambda \overset{ind}{\sim} \text{TP}((1 - \pi p)\lambda), \tag{11}$$

$$r_{ij} \mid s_{ij}, N_i, \pi, p, \lambda \overset{ind}{\sim} \text{Binomial} \left\{ s_{ij}, \frac{(1 - \pi)p}{1 - \pi p} \right\}, \tag{12}$$

$i = 1, \ldots, n$, $j = 1, \ldots, N_i - 1$. Thus, using the composition method, draws can be easily made from $\pi^{(ns)}(r^*, s^*, N \mid \pi, p, \lambda, r_0, s_0, a = 0)$.

In (a), we have

$$\pi_a^{(s)}(\pi, p \mid r_0, s_0, a) \propto \pi^{a-1}(1 - \pi)^{r-a-1}p^{r-1}(1 - p)^{s-r-1}H(\pi, p), 0 < \pi, p < 1,$$

where

$$H(\pi, p) = \int_0^\infty \frac{\lambda^{s-1} e^{-n\lambda}}{(1 - e^{-\lambda \pi p})^n} d\lambda.$$

One important question, "Does $H(\pi, p)$ exist?" arises. It can be shown that if $s \geq n + 1$ for $\pi, p > 0$, the integral does exist. The condition $s \geq n + 1$ is very weak because typically at least one cluster has more than one member. See Appendix A for a proof.

It is easy to show that the joint posterior density in (9) is proper. First, we note that the approximate joint posterior density $\pi_a^{(s)}(\pi, p \mid r_0, s_0, a)$ is proper provided that $n < a < r$ and $n < r < s$. This condition is easily satisfied. See Appendix B for a proof. Note again that $\{(1 - e^{-\lambda \pi p})/(1 - e^{-\lambda})\}^n$ is bounded; so clearly $T(r, N, \lambda, \pi, p)$ is bounded. Thus, because $\pi_a^{(s)}(\pi, p \mid r_0, s_0, a)$ is proper and $T(r, N, \lambda, \pi, p)$ is bounded, using the multiplication rule and (10), (11) and (12), it follows that the joint posterior density in (9) is proper.

Then, letting $s = \sum_{i=1}^n s_i$, $r = \sum_{i=1}^n r_i$, $a = \sum_{i=1}^n a_i$, the joint posterior density in (9) can be expressed as

$$\pi(r^*, s^*, N, \pi, p, \lambda \mid I, a, r_0, s_0) \propto$$

$$T(r, N, \lambda, \pi, p)\pi_a^{(ns)}(r^*, s^*, N \mid \pi, p, \lambda, r_0, s_j, a = 0)\pi_a^{(s)}(\pi, p, \lambda \mid r_0, s_0, a). \tag{13}$$

Note that

$$H(\pi, p) = E_T \left[1 - e^{-\pi p t/n} \right]^{-n},$$

where $T \sim \text{Gamma}(s, 1)$. Thus, a Monte Carlo estimator of $H(\pi, p)$ is

$$\hat{H}(\pi, p) \approx \frac{1}{M^*} \sum_{h=1}^{M^*} (1 - e^{-\pi p t^{(h)}})^{-n}, M^* \approx 10,000,$$

and $t^{(1)}, \ldots, t^{(M^*)} \overset{iid}{\sim} \text{Gamma}(s, 1)$. Note that we only need to draw this sample just once. We simply use a two-dimensional grid method to draw π and p. We have partitioned the unit square [i.e., $(0, 1) \times (0, 1)$] into 10,000 little squares (grids) of equal area, and we have approximated the joint posterior density by a joint probability mass function on these grids.

Next, we need to draw $\lambda \mid \pi, p, r_0, s_0, a$ from

$$P(\lambda \mid \pi, p, r_0, s_0, a) \propto \frac{\lambda^{s-1} e^{-n\lambda}}{(1 - e^{-\lambda \pi p})^n}, \lambda \geq 0.$$

Note that $P(\lambda \mid \pi, p, r_0, s_0, a)$ is not unimodal and, therefore, it is not log concave. To get unimodality, one would need λ to be much smaller than 1 which is unreasonable in our application because λ is approximately the expected sample size within a cluster. Then, standard accept-reject algorithms will not work efficiently and adaptive rejection sampling (Gilks and Wild[7]) is not available.

Again we can use a one-dimensional grid to draw λ and it is convenient to use a grid approximation for a density in (0, 1). We first make the transformation $\phi = \frac{\lambda}{\lambda+1}, 0 \le \phi \le 1$ to get

$$P(\phi \mid \pi, p, r_0, s_0, a) \propto \frac{1}{(1-\phi)^2} \left\{ \lambda^{s-1} \left(\frac{e^{-\lambda}}{1 - e^{-\lambda \pi p}} \right)^n \right\}_{\lambda = \frac{\phi}{1-\phi}}, 0 < \phi < 1.$$

We have approximated the conditional posterior density of ϕ by a discrete probability mass function of 100 equally spaced grid points in (0, 1). It is a good idea to use a grid method when it is not guaranteed that the density is unimodal. This method can be done in an adaptive manner; see Gelman et al.[6] (pg. 92). Note also that because we are not using Markov chain Monte Carlo methods, the issue of poor mixing (i.e., slow convergence) does not arise.

Thus, samples of $p^{(h)}, \pi^{(h)}, \lambda^{(h)}, N_i^{(h)}, s_{ij}^{(h)}, r_{ij}^{(h)}$, $h = 1, \ldots, M$, can be obtained by using the SIR algorithm to subsample the initial sample. The SIR algorithm works best when the weights have small variations around $1/\tilde{M}$, but this is difficult to achieve in complicated problems. Thus, it is best to run the SIR algorithm without replacement so that the values with the largest sampling weights are selected first; no value can be selected more than once because our original random sample of 10,0000 does not have repeated values. To make inference about R and P in (1), we substitute $N_i^{(h)}, s_{ij}^{(h)}, r_{ij}^{(h)}$ into R and P to get $P^{(h)}$ and $R^{(h)}$, $h = 1, \ldots, M$. Posterior inference is now available using standard Monte Carlo methods.

Under the ignorable selection model,

$$\pi^{(s)}(\pi, p, \lambda \mid r_0, s_0, a) \propto \frac{\lambda^{s-1} e^{-n\lambda} \pi^{a-1} (1 - \pi)^{r-a-1} p^{r-1} (1 - p)^{s-r-1}}{(1 - e^{-\lambda})^n},$$

and we can fit $\pi^{(s)}(\pi, p, \lambda \mid r_0, s_0, a)$ without using the SIR algorithm. We use a three-dimensional grid method to draw (π, p, λ), and we have transformed λ to $(0,1)$ using $\phi = \lambda/(1 + \lambda)$. We divide the ranges of the parameters π, p and λ into 100 grids to give 10^6 little cubes in three dimensions. Then we find the height of the posterior density at the centroid of each of these cubes (normalization constant is not needed). Thus, we can approximate the joint posterior by a discrete probability mass function with probabilities at the mass points proportional to the heights of the cubes. Thus, the SIR algorithm is not needed for fitting the ignorable selection model. Note that (b) is done in exactly the same manner as for the nonignorable selection model; here (10), (11) and (12) hold exactly. We have used $1,000$ draws for this method to make inference. For the ignorable selection model, the SIR algorithm gives virtually the same answers as the one without using the SIR algorithm.

We make a remark. Consider the approximate posterior density $\pi^{(s)}(p, \pi, \lambda \mid a, r_0, s_0)$. Because

$$\int_0^\infty \frac{\lambda^{s-1} e^{-n\lambda}}{[1 - e^{-\pi p \lambda}]^n} d\lambda \approx \frac{\Gamma(s-n)}{n^{s-n-1}} \frac{1}{(\pi p)^n},$$

$p \mid a, r_0, s_0$ and $\pi \mid a, r_0, s_0$ are independent with

$p \mid a, r_0, s_0 \sim \text{Beta}(r - n, s - r)$ and $\pi \mid a, r_0, s_0 \sim \text{Beta}(a - n, r - a)$.

Thus,

$$E(p \mid a, r_0, s_0) \approx \frac{r-n}{s-n} < \frac{r}{s} \text{ and } E(\pi \mid a, r_0, s_0) \approx \frac{a-n}{r-n} < \frac{a}{r},$$

where $\frac{r}{s}$ and $\frac{a}{r}$ are the corresponding posterior means under the ignorable selection model. The inequalities follow because $a < r < s$.

4. An Illustrative Example on Cystic Fibrosis

We discuss an example on cystic fibrosis in population genetics as we stated earlier. Cystic fibrosis is a rare disease and so it is inefficient to draw a random sample of families. One can view ascertainment as a sampling process in which unascertained families are discarded one by one until a family is finally ascertained. Families are ascertained through their affected siblings; see Lange.[10] If all siblings who first appear at a medical unit are not affected, their information is discarded and no more testing is done in that family; otherwise the entire family is tested. Ascertainment continues until another family is found with at least one affected sibling, and the process continues until the population is exhausted. The total number of affected members in the entire population and their proportion are of interest.

There are several methods to correct for ascertainment bias. With the exception of the recent method of Lange,[10] which uses sequential sampling, all methods are some variations of the method of Fisher[5] who assumed that there is a random sample which is corrected for ascertainment bias. These methods essentially estimate the segregation ratio (p) and proband probability (π), the primary parameters of interest. Specifically, see Nandram, Choi and Xu[16] for a review and a discussion of difficulties associated with maximum likelihood estimation for the ascertainment bias problem within the framework of Fisher.[5] However, researchers have begun to use Bayesian methods for the study of this problem; see, for example, Nandram and Xu.[18] It seems reasonable to include an intracluster correlation because the siblings in the same family might be linked by genetic traits, but Nandram and Xu[18] did not show substantial gain in doing so for these data. However, the approach of Nandram and Xu[18] is not particularly useful for finite population parameters like R and P. It is useful to note that our method is invariant to the order in which the families appear.

However, Lange[10] (Ch. 2) has a different perspective of this problem and he presented a sequential sampling scheme. He wrote "If we view ascertainment as a sampling process in which unascertained families of size s_k are discarded one by one until the kth ascertained family is finally ascertained, then the number of unascertained families follows a shifted geometric distribution ...." That is, although Lange[10] used a sequential sampling scheme, he did not study finite population quantities. This sequential scheme is attractive to estimate R and P.

We assume that the families are selected independently in a sequential process, and the probability that a family is selected is proportional to the family's number of affected siblings. That is, families with more affected siblings are more likely to be selected (i.e., one or more siblings are more likely to visit a medical unit). The total number of affected siblings and the proportion of affected siblings in the population are of interest. Previous research estimates the segregation probabilities and the proband probabilities; if the ascertainment bias is ignored, an estimate of these parameters will be inflated. Table 1 gives a set of data on cystic fibrosis, first presented by Crow[4] to illustrate the need to take account of the method of ascertainment in segregation analysis.

We have presented some comparisons in Table 2. We have compared the ignorable selection model and the nonignorable selection model when inferences are made about the segregation ratio (p), the proband probability (π), the expected family size (λ), the number of families (N), the finite population total at risk (R), the finite population total affected (R) and the finite population proportion affected (P). We have considered the cases in which the selection probabilities are proportional to the number at risk or the number affected in each family. While there are differences in these

Table 1. Number of families affected by sibship size, number affected offspring and number of probands (Crow[4]).

Size	Affected	Proband	Families	Size	Affected	Proband	Families
10	3	1	1	5	1	1	2
9	3	1	1	4	3	2	1
8	4	1	1	4	3	1	2
7	3	2	1	4	2	1	4
7	3	1	1	4	1	1	6
7	2	1	1	3	2	2	3
7	1	1	1	3	2	1	3
6	2	1	1	3	1	1	10
6	1	1	1	2	2	2	2
5	3	3	1	2	2	1	4
5	3	2	1	2	1	1	18
5	2	1	5	1	1	1	9

Note: Sibship sizes are different, ranging 1-10. The *probands* are the individuals in a family who first appear at the medical unit.

scenarios, the differences are small. The ignorable selection model is badly off whereas the nonignorable selection model provides a lot more reasonable answers. We have also compared Jeffreys' prior and a uniform prior for the estimation of these parameters, and as expected, there is very little difference.

We are basing our arguments for the nonignorable selection model on a well-known fact that for cystic fibrosis, p is expected to be .25 under Mendel's first law. So that if one of these two models gives the closest answer to .25, it is to be preferred. We use containment in the 95% credible intervals for p to support one of these models. It is not possible to prove mathematically or statistically that the nonignorable selection model is better than the ignorable selection model or any of the two models is, in fact, correct. The nonignorable selection model has an assumption that cannot be verified from the data: selection probabilities of the clusters are proportional to the r_{ij} or the s_{ij} and neither the selection probabilities or these two latter quantities are observed. Thus, standard Bayesian model selection procedures are not appropriate; see Section 6.7 of Gelman et al.[6] Indeed, as can be seen, this problem is difficult because of the lack of information.

We present Table 2. Under Jeffreys' prior the 95% credible interval for p is (.40, .52) under the ignorable selection model and (.23, .36) under the nonignorable selection model. Because (.23, .36) contains .25 and (.40, .52) is much above .25, there is credence in the nonignorable selection model, and with hindsight the nonignorable selection model is preferred over the ignor-

Table 2.   Comparison of the ignorable selection model (ISM) and the nonignorable selection model (NSM) for posterior inference about p, π, λ, N, S, R and P when selection probabilities are proportional to number at risk and number affected.

	PM	PSD	CI	PM	PSD	CI
	Jeffreys' prior			Uniform prior		
ISM						
p	0.46	0.03	(0.40, 0.52)	0.46	0.03	(0.40, 0.52)
π	0.73	0.04	(0.65, 0.80)	0.72	0.04	(0.65, 0.80)
λ	3.23	0.22	(2.85, 3.70)	3.23	0.22	(2.84, 3.69)
N	108	6	(96, 120)	108	6	(97 121)
S	337	18	(303, 373)	337	18	(305 373)
R	136	5	(128, 147)	137	5	(128 147)
P	0.41	0.02	(0.38, 0.44)	0.41	0.02	(0.37, 0.44)
NSM, at risk						
p	0.32	0.05	(0.23, 0.36)	0.34	0.03	(0.27, 0.36)
π	0.47	0.06	(0.35, 0.58)	0.51	0.05	(0.41, 0.61)
λ	2.56	0.18	(2.23, 2.94)	2.61	0.19	(2.24, 2.98)
N	172	16	(141, 201)	155	11	(133, 177)
S	494	42	(414, 576)	450	29	(399, 508)
R	170	14	(146, 199)	160	10	(139, 178)
P	0.35	0.03	(0.28, 0.38)	0.31	0.03	(0.26, 0.38)
NSM, affected						
p	0.30	0.05	(0.23, 0.36)	0.37	0.04	(0.27, 0.39)
π	0.49	0.05	(0.40, 0.58)	0.52	0.05	(0.45, 0.63)
λ	2.57	0.17	(2.23, 2.92)	2.62	0.19	(2.27, 3.00)
N	174	19	(139, 209)	157	13	(132, 184)
S	501	50	(406, 597)	456	35	(387, 519)
R	164	10	(146, 184)	157	8	(142, 173)
P	0.29	0.03	(0.25, 0.37)	0.35	0.03	(0.29, 0.39)

Note: PM is posterior mean, PSD is posterior standard deviation, and CI is the 95% credible interval. Here p is the segregation probability, π is the proband probability, and λ is the parameter corresponding to sibling size. Also N is the number of families at risk, S is the total number of siblings at risk, R is the total number of siblings affected, and $P = R/S$, the finite population proportion affected of the at risk population. Jeffreys' prior ($k = .5$) is used for λ in all models; for p and π, $k = .5$ for Jeffreys' prior and $k = 0$ for the uniform prior. The square of Jeffreys' prior gives virtually the same answers.

able selection model which inflates the estimate of p. Thus, we should rely on the nonignorable selection model for estimates of the more important parameters. Under the ignorable selection model the 95% credible intervals for N, S and R are respectively (96, 120), (303, 373) and (128, 147), and under the nonignorable selection model they are (141, 201), (414, 576) and (146, 199). It appears that the estimates under the ignorable selection model are much too small. Also, for both the ignorable and nonignorable selection models, these numbers compare very well with the uniform prior.

We also looked at the proportion of affected individuals relative to the number of individuals at risk, P. This takes the finite population into consideration. As can be seen, there are small differences between the posterior means of p and P, and as expected, there is a little bit more precision in estimating P.

5. Concluding Remarks

With limited amount of information, it is difficult to estimate the number of people affected, the number of people at risk, and therefore the proportion of people affected. Our Bayesian method with sequential modeling approach can help to estimate these parameters. We have used a simple hierarchical Bayesian model which captures a complex phenomenon. We have shown that inference can be made using Monte Carlo methods with random samples via the SIR algorithm rather than samples from Markov chains. This is a desirable simplification of the Bayesian method.

Using an example of cystic fibrosis, we have shown that with Jeffreys' prior the 95% credible interval for p is (.40, .52) under the ignorable selection model and (.23, .36) under the nonignorable selection model. We have argued that because (.23, .36) contains .25 and (.40, .52) is much above .25, there is credence in the nonignorable selection model. Otherwise, it is impossible to tell which of these two models is better. However, in the nonignorable selection model it does not matter whether the unknown selection probabilities are proportional to the sizes of the clusters or the number with the characteristic; one does not need to choose one over the other. That is, this unverifiable assumption is not too restrictive and our methodology provides an opportunity to do some sensible analysis with these data. However, further testing of our nonignorable selection model will be useful.

While we have used an illustrative example on cystic fibrosis, our methodology has wider generality in survey sampling. The survey sampling design is important when it is difficult to get samples from the population with the desired characteristics. Thus, in the realm of survey sampling our model can be improved in several ways. It is possible to include more informative prior information into our model and in other applications it will be useful to include an intracluster correlation. The models of Nandram and Xu[18] include the intracluster correlation, but these models are not used for predicting finite population quantities under selection bias, a much more difficult problem.

First, the data on such applications are usually sparse and the occurrence of a positive individual is usually rare, thereby making prediction difficult. Therefore, any useful information about the parameters p and λ can be incorporated a priori. For example, a family size is expected to be about 2 and not much larger than 5. Therefore, $E(\lambda) = \mu_0 \approx 2$ and a 95% upper confidence bound of λ is $b_o \approx 5$. Thus, we take $\lambda \sim \text{Gamma}(\alpha, \beta)$, where $\beta = G_{\mu_o \beta}^{-1}(0.95)/b_o$, $G^{-1}(\cdot)$ is the inverse cumulative distribution function of the gamma random variable and $\alpha = \mu_o \beta$. With a simple iterative procedure one gets $\lambda \sim \text{Gamma}(1.70, 0.85)$. Again, this is a useful piece of prior information.

Second, in a more general setting we can incorporate an intracluster correlation. This correlation can be included using a binomial model (e.g., Altham[1]). Recent applications of this model for contingency tables are given by Nandram and Choi[13,14]. The intracluster correlation, $\theta > 0$, can be added to (3), $r_{ij} \mid p, s_{ij}, N_i \overset{ind}{\sim} \text{Binomial}(s_{ij}, p), j = 1, \ldots, N_i, i = 1, \ldots, n$. Altham's formula then gives $P(r_{ij} \mid p, s_{ij}, N_i) = \theta(1 - p) + (1 - \theta)(1 - p)^{s_{ij}}, r_{ij} = 0, s_{ij} \geq 1$; $P(r_{ij} \mid p, s_{ij}, N_i) = \theta p + (1 - \theta)p^{s_{ij}}, r_{ij} = s_{ij} \geq 1$; and $P(r_{ij} \mid p, s_{ij}, N_i) = \theta(1-p)\{s_{ij}!/r_{ij}!(s_{ij} - r_{ij})!\}p^{r_{ij}}(1-p)^{s_{ij}-r_{ij}}, r_{ij} = 1, \ldots, s_{ij} - 1, s_{ij} \geq 2$.

Finally, we discuss another application of considerable interest. A city map is partitioned into grids (clusters); the number of nonempty grids is unknown. We want to estimate the number of people with a certain characteristic (e.g., homeless people, strangers passing through the city, tourists, etc.). We sample these grids sequentially, each in turn, and the order in which the grids are sampled is irrelevant because all the grids will be sampled. Once the sequentially sampling comes to a grid, at least one individual is sampled and interviewed. If none of the individuals has the characteristic, the grid is discarded and no information is recorded. Thus, there is a selection bias because there may be individuals with the characteristic in the nonsampled part of the grid. However, if at least one individual has the characteristic, all individuals in this grid are interviewed and the information is recorded. Sequential sampling continues until all grids are visited. In an emergency, the grids can be divided into groups and independent sequential sampling can be done. We need to estimate the number of people with the characteristic of interest.

Acknowledgments

We are grateful to the two referees who helped to improve the presentation enormously.

APPENDIX A: Existence of $H(\pi, p)$

We show that $\int_0^\infty \frac{\lambda^{s-1} e^{-n\lambda}}{(1-e^{-\pi p\lambda})^n} d\lambda$ is finite.
First, we note that

$$\int_0^\infty \frac{\lambda^{s-1} e^{-n\lambda}}{(1 - e^{-\pi p\lambda})^n} d\lambda = \int_0^1 \frac{\lambda^{s-1} e^{-n\lambda}}{(1 - e^{-\pi p\lambda})^n} d\lambda + \int_1^\infty \frac{\lambda^{s-1} e^{-n\lambda}}{(1 - e^{-\pi p\lambda})^n} d\lambda. \quad (A.1)$$

Because $\pi p\lambda \geq \pi p$ on $(1, \infty)$,

$$\int_1^\infty \frac{\lambda^{s-1} e^{-n\lambda}}{(1 - e^{-\pi p\lambda})^n} d\lambda < \frac{1}{(1 - e^{-\pi p})^n}.$$

Now, since $e^{-\pi p} \geq 1 - \pi p$,

$$\int_1^\infty \frac{\lambda^{s-1} e^{-n\lambda}}{(1 - e^{-\pi p\lambda})^n} d\lambda \leq (\pi p)^{-n} A_1, \quad (A.2)$$

where $A_1 = \int_1^\infty \lambda^{s-1} e^{-n\lambda} d\lambda < \infty$.

Because $n\lambda \geq n\pi p\lambda$, $\int_1^\infty \frac{\lambda^{s-1} e^{-n\lambda}}{(1-e^{-\pi p\lambda})^n} d\lambda \leq \int_0^1 \lambda^{s-1} (1 - e^{-\pi p\lambda})^{-n} d\lambda$. Now, using the fact that $e^{\pi p\lambda} \geq 1 + \pi p\lambda$, we have

$$\int_1^\infty \frac{\lambda^{s-1} e^{-n\lambda}}{(1 - e^{-\pi p\lambda})^n} d\lambda \leq (\pi p)^{-n} \int_0^1 \lambda^{s-n-1} d\lambda = (\pi p)^{-n}/(s - n), \quad (A.3)$$

provided that $s \geq n + 1$.

Thus, by (A.2) and (A.3),

$$H(\pi, p) \leq (\pi p)^{-n} A, \quad \text{where } A < \infty. \quad (A.4)$$

Thus, $H(\pi, p) < \infty$ provided $\pi, p > 0$ and $s \geq n+1$. Note that the condition, $s \geq n + 1$, is satisfied if there is one family with at least two siblings; this is a very weak condition because it will be satisfied almost surely.

APPENDIX B: Propriety of $\pi_a^{(s)}(\pi, p \mid \underset{\sim}{r}_0, \underset{\sim}{s}_0, \underset{\sim}{a})$

In Appendix A, we have shown that $H(\pi, p) < \infty$ provided $\pi, p > 0$ and $s \geq n + 1$. Thus, we only need to show that

$$C = \int_0^1 \int_0^1 \frac{1}{(\pi p)^n} \pi^{a-1} (1 - \pi)^{r-a-1} p^{r-1} (1 - p)^{s-r-1} d\pi dp < \infty. \quad (B.1)$$

But C in (B.1) is a product of two integrals because

$$C = \int_0^1 \pi^{a-n-1} (1 - \pi)^{r-a-1} d\pi \int_0^1 p^{r-n-1} (1 - p)^{s-r-1} dp. \quad (B.2)$$

In (B.2) the first integral is finite if $n < a < r$, and the second integral is finite if $n < r < s$. Both conditions are satisfied.

References

1. P. M. E. Altham, *Biometrika* **63**, 263 (1976).
2. R. Chambers, A. Dorfman, and S. Wang, *Journal of the Royal Statistical Society, Series B* **60**, 397 (1998).
3. W. G. Cochran, W. G. *Sampling Techniques*, 3rd edition (Wiley, New York, 1977).
4. J. F. Crow, *Public Health Service Publication 1163*, in *Epidemiology and Genetics of Chronic disease*, Eds: Neal, J. V., Shaw, M. W., Schull, W. J., Department of Health, Education, and welfare, Washington, DC, 23 (1965).
5. R. A. Fisher, *Annals of Eugenics* **6**, 13 (1934).
6. A. Gelman, J. B. Carlin, H. S. Stern and D. B. Rubin, *Bayesian Data Analysis*, 2nd edition (Chapman & Hall/CRC, New York, 2004).
7. W. R. Gilks and P. Wild, *Applied Statistics* **41**, 337 (1992).
8. P. J. Green, *Biometrika* **82**, 711 (1995).
9. A. Krieger and D. Pfeffermann, *Survey Methodology* **18**, 225 (1992).
10. K. Lange, *Mathematical and Statistical Methods for Genetic Analysis*, 2nd edition (Springer-Verlag, New York, 2002).
11. D. Malec, W. W. Davis and X. Cao, *Statistics in Medicine* **18**, 3189 (1999).
12. B. Nandram, Bayesian Predictive Inference under Informative Sampling via Surrogate Sampling, in *Bayesian Statistics and Its Applications*, Eds. S.K. Upadhyay, Umesh Singh and Dipak K. Dey, Anamaya, New Delhi, Chapter **25**, 356 (2007).
13. B. Nandram and J. W. Choi, *Journal of Statistical Computation and Simulation* **76**, 233 (2006).
14. B. Nandram and J. W. Choi, *Journal of Data Science* **5**, 217 (2007).
15. B. Nandram and J. W. Choi, *Journal of the American Statistical Association* **105**, 120 (2010).
16. B. Nandram, J. W. Choi and H. Xu, *Journal of Data Science* **9**, 23 (2011).
17. B. Nandram, J. W. Choi, G. Shen and C. Burgos, *Applied Stochastic Models in Business and Industry* **22**, 559 (2006).
18. B. Nandram and H. Xu, *Journal of Biometrics and Biostatistics* **2**, 112 (2011).
19. A. E. Raftery, *Biometrika* **75**, 223 (1988).
20. C. P. Robert and G. Casella (1999), *Monte carlo Statistical Methods* (Springer, New York, 1999).
21. A. F. M. Smith and A. E. Gelfand, *American Statistician* **46**, 84 (1992).

TESTS FOR AND AGAINST UNIFORM STOCHASTIC ORDERING ON MULTINOMIAL PARAMETERS BASED ON ϕ-DIVERGENCES

JIANAN PENG

Department of Mathematics and Statistics, Acadia University,
Wolfville, NS B4P 2R6, Canada
E-mail: jianan.peng@acadiau.ca

Measures of phi-divergences are widely used as statistical tools for inferences about multinomial parameters and more general than the famous Cressie and Read's power divergence. In this article, tests based on measures of phi-divergence are considered for and against uniform stochastic ordering on multinomial parameters. The asymptotic distributions of the test statistics are shown to be of the chi-bar-squared type. An example is used to illustrate the method developed.

1. Introduction

Pearson's chi-square and likelihood ratio are two of the most popular test statistics of testing equality of a multinomial probability vector to a given probability vector. Cressie and Read[3,10] introduced the famous family of power divergence test statistics. For two probability vectors $\mathbf{p} = (p_1, \ldots, p_k)$ and $\mathbf{q} = (q_1, \ldots, q_k)$, the power divergence family of test statistics is denoted by $2nI^\lambda(\hat{\mathbf{p}}, \mathbf{q})$, $\hat{\mathbf{p}}$ having been determined from a random sample of size n, where

$$I^\lambda(\mathbf{p}, \mathbf{q}) = \frac{1}{\lambda(\lambda + 1)} \sum_{i=1}^{k} p_i \left[\left(\frac{p_i}{q_i} \right)^\lambda - 1 \right] \tag{1}$$

with the cases of $\lambda = 0, -1$ defined as the continuous limits at those values of λ. It can be easily seen that the statistics such as Neyman modified chi-square, minimum discriminant information, Freeman-Tukey, log likelihood ratio and the Pearson's chi-square are special cases of the power divergence statistic with $\lambda = -2, -1, -\frac{1}{2}, 0, 1$ values, respectively. Cressie and Read (*loc. cit.*) studied the differences in behavior of (1) asymptotically and for finite sample sizes for different λ values. They suggested using the statistic based on $\lambda = \frac{2}{3}$ as a competitor to the Pearson's chi-square and log likelihood ratio statistics. Notice that Cressie and Read's power divergence family is the special case of the Csiszar ϕ-divergence family introduced by Csiszar[2] and Ali and Silvey.[1] The Csiszar ϕ-divergence family between $\mathbf{p}$ and $\mathbf{q}$ is defined by $D_\phi(\mathbf{p}, \mathbf{q}) = \sum_{i=1}^{k} q_i \phi(\frac{p_i}{q_i})$ for every convex function $\phi : [0, \infty) \to R \cup \{\infty\}$, where $0\phi(\frac{0}{0}) = 0$ and $0\phi(\frac{p}{0}) = \lim_{u \to \infty} \frac{\phi(u)}{u}$. When $\phi_\lambda(x) = \frac{(x^{\lambda+1} - x)}{\lambda(\lambda+1)}, \lambda \neq 0, -1$, we have the power divergence family. When $\phi(x) = \frac{(x-1)^2}{2}$, we have the Pearson' chi-squared statistic. For more examples of ϕ, see Pardo.[8] Even though the Csiszar ϕ-divergence family is very general, it fails to include some popular measures of divergence that cannot be written as ϕ-divergences. These include measures by Renyi, Bhattacharya, and Sharma-Mittal. However, these three measures can be written as continuous functions of strictly increasing functions of some specific ϕ-divergences (see Section 3).

In many situations prior information regarding the specific nature of the p_i's in a multinomial model is available, say, they are nondecreasing. Incorporating such information within statistical analysis leads to more efficient procedures (see Robertson et al.[11] and Silvapulle and Sen[13] and references therein). Stochastic orderings are used to compare probability distributions. Let $\mathbf{p}$ and $\mathbf{q}$ be two multinomial parameters. The vector $\mathbf{p}$ is said to be larger that $\mathbf{q}$ in, (i) the stochastic order, denoted by $\mathbf{p} \geq_{st} \mathbf{q}$, if $\sum_{j=i}^{k} p_j \geq \sum_{j=i}^{k} q_j$, for $i = 1, 2, \ldots, k$; (ii) the uniform stochastic order or hazard ratio order, denoted by $\mathbf{p} \geq_u \mathbf{q}$, if $\sum_{j=i}^{k} p_j / \sum_{j=i}^{k} q_j$ is nondecreasing in i; and (iii) the likelihood ratio order, denoted by $\mathbf{p} \geq_{lr} \mathbf{q}$, if p_i/q_i is nondecreasing in i. It is well known that the likelihood ratio ordering implies the uniform stochastic ordering, which in turn implies the stochastic ordering (see Shaked and Shanthikumar[12] and Lee et al.[5]).

Although the likelihood ratio test is the most common choice as a test statistic in existing order restricted testing problems, many of the other divergence test statistics have better power under certain alternatives. For example, Menendez et al.[7] demonstrated this when the alternative is simple ordered. Lee and Yan[6] showed this when the alternative is under the uniform stochastic ordering. Cressie and Read,[3] and Read and Cressie[10] demonstrated similar cases when the alternative is unrestricted.

We consider the following three hypotheses $H_0 : \mathbf{p} = \mathbf{q}, H_1 : \mathbf{p} \geq_u \mathbf{q}$, and $H_2 :$ No restriction, where $\mathbf{q}$ is a known probability vector. We assume that $p_i > 0, i = 1, \ldots, k$ and require a large sample size so that $\hat{p}_i = \frac{n_i}{n} >$

$0, i = 1, \ldots, k$, where n_i denotes the observed frequency for the ith category in a multinomial experiment and $\sum_{i=1}^{k} n_i = n$. Park et al.[9] considered the classical likelihood ratio test for H_1 versus $H_2 - H_1$. As shown in Lee and Yan,[6] Cressie and Read's power divergence tests in this case may be more powerful. In this article, we consider some families of test statistics for testing H_0 versus $H_1 - H_0$ and H_1 versus $H_2 - H_1$ based on ϕ-divergences as defined in Section 3. These families of test statistics include as a special case, the power divergence family in Lee and Yan.[6] In Section 2, the maximum likelihood estimator of $\mathbf{p}$ under the uniform stochastic ordering restriction is introduced. In Section 3, the ϕ−divergence statistics for and against the uniform stochastic ordering are given and their asymptotic distributions are proved to be of the chi-bar-squared type and do not depend on the function of ϕ. In Section 4, a numerical example is discussed to illustrate the procedures developed.

2. Maximum Likelihood Estimator P* of P under P $\geq_U$ Q

It can be shown that the maximum likelihood estimation of $\mathbf{p}$ under the restriction $\mathbf{p} \geq_u \mathbf{q}$ is equivalent to minimize $\sum_{i=1}^{k} p_i ln \frac{p_i}{q_i}$ with respect to p under the restriction $\mathbf{p} \geq_u \mathbf{q}$. Notice that the constraint $\mathbf{p} \geq_u \mathbf{q}$ requires that $\frac{\sum_{j=i}^{k} p_j}{\sum_{j=i}^{k} q_j}$ is nondecreasing in i or equivalently,

$$\frac{p_i}{\sum_{j=i+1}^{k} p_j} \leq \frac{q_i}{\sum_{j=i+1}^{k} q_j}, i = 1, 2, \ldots, k - 1.$$

We reparametrize $\mathbf{p}$ by letting $\theta_i = \frac{p_i}{\sum_{j=i+1}^{k} p_j}, i = 1, 2, \ldots, k - 1$. Then $p_i = \frac{\theta_i}{\tau_i}, i = 1, 2, \ldots, k$, where $\tau_i = \prod_{j=1}^{i}(1 + \theta_j), i = 1, 2, \ldots, k - 1$. We use the convention $\theta_k = 1, \tau_k = \tau_{k-1}$. Similarly, We define $\hat{\theta}_i$ and $\hat{\tau}_i$ for $\hat{\mathbf{p}}$ as $\hat{\theta}_i = \frac{\hat{p}_i}{\sum_{j=i+1}^{k} \hat{p}_j}, \hat{\tau}_i = \prod_{j=1}^{i}(1 + \hat{\theta}_j)$ and ξ_i for $\mathbf{q}$ as $\xi_i = \frac{q_i}{\sum_{j=i+1}^{k} q_j}$ with the same convention $\hat{\theta}_k = 1, \hat{\tau}_k = \hat{\tau}_{k-1}$ and $\xi_k = 1$. After the reparametrization, the minimization problem $\min_{\mathbf{p} \geq_u \mathbf{q}} \sum_{i=1}^{k} \hat{p}_i ln \frac{\hat{p}_i}{q_i}$ becomes the problem of minimizing $\sum_{i=1}^{k} \frac{\hat{\theta}_i}{\hat{\tau}_i} ln \frac{\hat{\theta}_i \tau_i}{\hat{\tau}_i \theta_i}$ subject to the restriction $\theta_i \leq \xi_i, i = 1, \ldots, k - 1$. The following result is a special case in Lee and Yan[6] for $\lambda = 0$.

Proposition 2.1. Let $\mathbf{p}^*$ be the solution that minimizes $\sum_{i=1}^{k} p_i \ln \frac{p_i}{\hat{p}_i}$ subject to the restriction

$$\frac{p_i}{\sum_{j=i+1}^{k} p_j} \leq \frac{q_i}{\sum_{j=i+1}^{k} q_j}, i = 1, \ldots, k - 1.$$

Then

$$p_i^* = \frac{\hat{\theta}_i^*}{\prod_{j=1}^i (1 + \hat{\theta}_j^*)}, i = 1, \ldots, k-1; p_k^* = \frac{1}{\prod_{j=1}^{k-1}(1 + \hat{\theta}_j^*)},$$

where

$$\hat{\theta}_i^* = \min(\hat{\theta}_i, \xi_i), i = 1, 2, \ldots, k-1.$$

3. Test Statistics Based on Φ-Divergence

Based on Cressie and Read's family of power divergence statistics, Lee and Yan[6] used $T_{01}^\lambda = 2n[I^\lambda(\hat{\mathbf{p}}, \mathbf{q}) - I^\lambda(\hat{\mathbf{p}}, \mathbf{p}^*)]$ and $T_{12}^\lambda = 2nI^\lambda(\hat{\mathbf{p}}, \mathbf{p}^*)$ for testing H_0 versus $H_1 - H_0$ and H_1 versus $H_2 - H_1$, respectively. By Jensen's inequality it follows that $I^\lambda(\mathbf{p}, \mathbf{q})$ is always non-negative and it is zero if and only if $\mathbf{p} = \mathbf{q}$. Furthermore, if each $p_i > 0$, $I^\lambda(\mathbf{p}, \mathbf{q})$ is a strictly convex function of $\mathbf{q}$ over the set of probability vectors. Therefore, $I^\lambda(\mathbf{p}, \mathbf{q})$ is a discrepancy measure between $\mathbf{p}$ and $\mathbf{q}$, although it should be noted that $I^\lambda(\mathbf{p}, \mathbf{q})$ is not a metric.

But there is a more general family of statistics that includes the above family of power divergence statistics, namely,

$$T_{01}^\phi = \frac{2n}{\phi''(1)} (D_\phi(\hat{\mathbf{p}}, \mathbf{q}) - D_\phi(\hat{\mathbf{p}}, \mathbf{p}^*)) \tag{2}$$

and

$$T_{12}^\phi = \frac{2n}{\phi''(1)} D_\phi(\hat{\mathbf{p}}, \mathbf{p}^*). \tag{3}$$

The limiting distribution T_{01}^ϕ under H_0 and the limiting distribution of T_{12}^ϕ at the least favorable null parameter $\mathbf{p}$ are given below.

Theorem 3.1. *If $H_0 : \mathbf{p} = \mathbf{q}$ is true and $\phi(x)$ is twice continuously differentiable in the neighborhood of $x = 1$, with $\phi(1) = \phi'(1) = 0, \phi''(1) > 0$, then*

$$\lim_{n \to \infty} P(T_{01}^\phi \le t) = \sum_{i=0}^{k-1} \binom{k-1}{i} P(\chi_i^2 \le t)/2^{k-1} \tag{4}$$

and

$$\lim_{n \to \infty} \sup_{\mathbf{p} \ge_u \mathbf{q}} P(T_{12}^\phi \le t) = \sum_{i=0}^{k-1} \binom{k-1}{i} P(\chi_i^2 \le t)/2^{k-1}, \tag{5}$$

for real $t > 0$, where the convention $\chi_0^2 = 0$ is used.

Proof. The theorem holds by using Lemma 3.1 below and Theorem 2.1 in Lee and Yan.[6] Note that Pearson's chi-squared test statistic is a special case of $\lambda = 1$ in Lee and Yan.[6] □

Note. The approximate critical values for the above limit distributions for $k = 3, 4, \ldots, 15$ and $\alpha = 0.10, 0.05, 0.01$ are in given in Table 5.3.1 in Robertson *et al.*[11]

Lemma 3.1. *For each $\phi(x)$ satisfying the conditions in Theorem 3.1, $T_{01}^{\phi} = T_{01}^1 + o_p(1)$ and $T_{12}^{\phi} = T_{12}^1 + o_p(1)$, where T_{01}^1 and T_{12}^1 are the Pearson's Chi-squared test statistics in Lee and Yan[6] for $\lambda = 1$.*

Proof of Lemma 3.1.

Similar to the proof of Theorem 2.1 in Menendez *et al.*,[7] a second order Taylor expansion of $f(x) = x\phi(\frac{a}{x})$ around $\hat{p}_i$, where $x = p_i^*$ and $a = \hat{p}_i$, gives $T_{12}^{\phi} = \frac{2n}{\phi''(1)} \sum_{i=1}^{k} p_i^* \phi(\frac{\hat{p}_i}{p_i^*}) = \sum_{i=1}^{k} \frac{(n\hat{p}_i - np_i^*)^2}{np_i^*} + o_p(1)$.

Theorem 3.1 in $Pardo$[8] has proved that $\frac{2n}{\phi''(1)} D_{\phi}(\hat{\mathbf{p}}, \mathbf{q})$ is asymptotically equivalent to $\sum_{i=1}^{k} \frac{(n\hat{p}_i - nq_i)^2}{nq_i}$. Using expressions (2) and (3) and the result we have proved for T_{12}^{ϕ}, T_{01}^{ϕ} is equivalent to $2nI^1(\hat{\mathbf{p}}, \mathbf{q}) - 2nI^1(\hat{\mathbf{p}}, \mathbf{p}^*) = T_{01}^1$ in Lee and Yan.[6] □

There are some important measures of divergence such as those given by Batthacharya, Renyi and Sharma-Mittal (see Pardo[8]), which cannot be written as ϕ-divergences. However, these measures can be written in the following way: $D_{\phi,h} = h(D_{\phi}(\mathbf{p}, \mathbf{q}))$, where h is an increasing differentiable function from $[0, \infty)$ to $[0, \infty)$ and $h(0) = 0, h'(0) > 0$. The following table lists these divergence measures. In the case of Rényi's diver-

Table 1. Examples of divergence measures that can't be written as ϕ-divergences.

Divergence	$h(x)$	$\phi(x)$
Battacharya	$-\log(-x + 1)$	$-x^{\frac{1}{2}} + \frac{1}{2}(x+1)$
Renyi	$\frac{\log(1+r(r-1)x)}{r(r-1)}, r \neq 0, 1$	$\frac{x^r - r(x-1)-1}{r(r-1)}, r \neq 0, 1$
Sharma-Mittal	$\frac{(1+r(r-1)x)^{\frac{\nu-1}{r-1}}-1}{\nu-1}, \nu, r \neq 1$	$\frac{x^r - r(x-1)-1}{r(r-1)}, r \neq 0, 1$

gence measure, we have $D_{Renyi}^r(\mathbf{p}, \mathbf{q}) = \frac{1}{r(r-1)} \log(\sum_{i=1}^{k} p_i^r q_i^{1-r}), r \neq 0, 1$.

For $r = 0$ and 1, we define $D^r_{Renyi}(\mathbf{p}, \mathbf{q})$ by continuity in r and obtain $D^0_{Renyi}(\mathbf{p}, \mathbf{q}) = lim_{r \to 0} D^r_{Renyi}(\mathbf{p}, \mathbf{q}) = \sum_{i=1}^{k} q_i \log \frac{q_i}{p_i}$ and $D^1_{Renyi}(\mathbf{p}, \mathbf{q}) = lim_{r \to 1} D^r_{Renyi}(\mathbf{p}, \mathbf{q}) = \sum_{i=1}^{k} p_i \log \frac{p_i}{q_i}$. We can define a family of statistics $T^{\phi,h}_{01}$ and $T^{\phi,h}_{12}$, by

$$T^{\phi,h}_{01} = \frac{2n}{\phi''(1)h'(0)}(D_{\phi,h}(\hat{\mathbf{p}}, \mathbf{q}) - D_{\phi,h}(\hat{\mathbf{p}}, \mathbf{p}^*))$$

and

$$T^{\phi,h}_{12} = \frac{2n}{\phi''(1)h'(0)}D_{\phi,h}(\hat{\mathbf{p}}, \mathbf{p}^*),$$

which has the same asymptotic distribution as T^{ϕ}_{01} and T^{ϕ}_{12}, respectively. In the particular case of Rényi's divergence we have $h'(0) = 1$ and $\phi''(1) = 1$. Then

$$T^{Renyi,r}_{01} = 2n(D^r_{Renyi}(\hat{\mathbf{p}}, \mathbf{q}) - D^r_{Renyi}(\hat{\mathbf{p}}, \mathbf{p}^*)) \qquad (6)$$

and

$$T^{Renyi,r}_{12} = 2nD^r_{Renyi}(\hat{\mathbf{p}}, \mathbf{p}^*). \qquad (7)$$

4. An Example

In this section we use the data set discussed in a report from the Boston Collaborative Drug Surveillance Program (1974) to illustrate the new statistics introduced in this article. This data set has been analyzed by Dykstra et al.[4] and Lee and Yan[6] for different purposes. The data set consists of measurements for the mean daily insulin dose from 80 subjects identified as "hypoglycemia present" and 245 subjects as "hypoglycemia absent". The measurements are classified into five categories and are presented in Table 2. Let $\mathbf{p}$ be the parameters for the population with hypoglycemia and $\mathbf{q}$ be the parameters for the population without hypoglycemia, respectively. Since the sample size in the case of "hypoglycemia absent" is quite large, we assume that $\hat{\mathbf{q}} = q$ for illustrative purpose, as Lee and Yan[6] did. It is known that hypoglycemia (low blood sugar) would occur when large amounts of glucose are metabolized and hence would be consistent with a higher level of insulin dosage. Table 2 also contains the maximum likelihood estimate $\mathbf{p}^*$ of $\mathbf{p}$ under the restriction $\mathbf{p} \geq_u \mathbf{q}$.

The ϕ-divergence-based test statistics for testing H_0 versus $H_1 - H_0$ and H_1 versus $H_2 - H_1$ for selected values of $r = -2, -0.5, \frac{2}{3}, 2, 3$ are given in Table 3. Here, I^r stands for the power divergence family. R^r stands for the Rényi divergence family defined in (6) and (7). As in Lee and Yan,[6] the

Table 2. Mean daily insulin dose and unrestricted estimates.

	Insulin Level				
	<2.5	0.25-0.49	0.50-0.74	0.75-0.99	≥ 1.0
Hypo. Present	4	21	28	15	12
Hypo. Absent	40	74	59	26	46
$\hat{p}$	0.0500	0.2625	0.3500	0.1875	0.1500
$\mathbf{q}$	0.1633	0.3020	0.2408	0.1061	0.1878
$\hat{\theta}$	0.0526	0.3818	1.0370	1.2500	
ϕ	0.1951	0.5649	0.8194	0.5652	
$\hat{\theta}^*$	.0526	.3818	.8194	.5652	
p^*	.0500	.2625	.3096	.1365	.2414

p-values were obtained via the Monte Carlo simulation with 10000 replications. The random samples were generated from the multinomial distribution with the probability vector $(0.1633, 0.3020, 0.2408, 0.1061, 0.1878)$ and $n = 80$. The simulated p-values match very well with the results in Lee and Yan[6] for the power divergence family. One could also use Theorem 3.1 to obtain the asymptotic p-values. The simulated p-values are preferable to the asymptotic p-values but they are close to each other. The p-values for testing H_0 versus $H_1 - H_0$ are all less than 0.01 significance level. The null hypothesis $H_0 : \mathbf{p} = \mathbf{q}$ is therefore rejected in favor of $H_1 : \mathbf{p} \geq_u \mathbf{q}$. The Renyi divergence family for $r = 2/3$ is more powerful than the Cressie-Read power divergence for $r = 2/3$ when testing H_0 versus $H_1 - H_0$. All the p-values for testing H_1 versus $H_2 - H_1$ are larger than 0.05 significance level and therefore, there is insufficient evidence to reject H_1 at $\alpha = 0.05$.

Table 3. Testing statistics and p-values for the insulin dose data.

r	$T_{01}^{I^r}$	$T_{12}^{I^r}$	$T_{01}^{R^r}$	$T_{12}^{R^r}$
-2	21.381(.0052)	5.948 (.0781)	19.315 (.0096)	6.000 (.0703)
-0.5	13.327 (.0039)	5.160 (.0915)	16.462 (.0042)	5.557 (.0820)
2/3	11.634 (.0048)	4.795 (.1052)	13.141 (.0039)	5.106 (.0938)
2	12.049 (.0061)	4.583 (.1188)	10.228 (.0061)	4.581 (.1079)
3	13.585 (.0068)	4.545 (.1245)	8.704 (.0099)	4.221 (.1195)

Acknowledgments

This work was supported in part by the Natural Sciences and Engineering Research Council of Canada. The author is grateful to Editor Dr. Yogendra Chaubey and two referees for their comments which improved this article.

References

1. S. M. Ali and S. D. Silvcy, *J. Roy. Statist. Soc., Series B* **28**, 131 (1966).
2. I. Csiszar, *Publications of the Mathematical Institute of Hungarian Academy of Sciences, Series A* **8**, 84 (1963).
3. N. Cressie, T. R. C. Read, *J. Roy. Statist. Soc. B* **46**, 440 (1984).
4. R. L. Dykstra, S. Kochar, T. Robertson, *J. Amer. Statist. Assoc.* **90**, 1034 (1995).
5. C. Lee, C. Park, J. Peng, X. Yan, *Adv. Appl. Stat. Sci* **3**, 65 (2010).
6. C. I. C. Lee and X. Yan, *J. Statist. Plann. Inference* **107**, 267 (2002).
7. M. L. Menendez, L. Pardo and K. Zografos, *Utilitas Mathematica* **61**, 209 (2002).
8. L. Pardo, *Statistical Inference Based on Divergence Measures* (Chapman and Hall/CRC, 2006).
9. C. G. Park, C. C. Lee and T. Robertson, *Canad. J. Statist* **26**, 69 (1998).
10. T. R. C. Read, N. Cressie, *Goodness of fit for discrete multivariate data* (Springer, New York, 1988).
11. T. Robertson, F. T. Wright, and R. L. Dykstra, *Ordered Restricted Statistical Inference* (Wiley, Chichester, 1988).
12. M. Shaked, J. G. Shanthikumar, *Stochastic Orders and Their Applications* (Academic Press, Boston, 1994).
13. M. J. Silvapulle and P. K. Sen, *Constrained Statistical Inference* (Wiley, New Jersey, 2005).

DEVELOPMENT AND MANAGEMENT OF NATIONAL HEALTH PLANS: HEALTH ECONOMICS AND STATISTICAL PERSPECTIVES

PRANAB KUMAR SEN

Departments of Biostatistics, and Statistics & Operations Research,
University of North Carolina, Chapel Hill, NC 27599-7420, USA
E-mail: pksen@bios.unc.edu

National health plans should pertain to all geo-politic and socio - economic strata of the nation and be compatible with available national resources in compliance with the diverse need of the national health and welfare system. For this highly complex multi-factor setup, incorporation of statistical rationality and interpretation is essential. Many qualitative socio-economic measures are needed to be expressed in a quantitative way. Some of these measures are statistically appraised in the light of health economics.

1. Introduction

There is an innumerable number of factors, not all causal nor all working reconcilably, which undermine the complex of national health spectrum, its relation to global health, its diversity in the inclusion of people from all walks of society, its support by a matching national health economics plan, its impact on human/subhuman *quality of life* (QOL), and its aim to avert eco-environmental imbalance to disasters. For such high-dimensional models there is a genuine need for an interdisciplinary approach wherein health economics, socio-economics and geo-politics are to be properly blended with information and bio-technology, genetics, clinical and public health sciences, and statistical science. It is essential to comprehend statistically *knowledge discovery and data mining* (KDDM) tools for valid data collection, monitoring and interpretation of this complex model, with vivid local, regional, as well as, global perspectives. The task is appealing but challenging; an accomplishment is rewarding. Cumulation of persistent inequality of income (wealth), education, health care, life style and outlook on life may wipe out all the so called economic progress and prosperity. It is necessary to incorporate statistical interpretation and rationality in many socio-economic

and health measures to assess the real depth of this complex problems and
the impacted impasses.

2. An Inventory of Points to Ponder

(1) Poverty - affluence and urban - rural differentials in health care.

(2) Health care facilities: Private vs public entities.

(3) BPL, AAL and the vanishing middle class.

(4) Social security, medicare and health economics.

(5) Occidental vs oriental medical practice.

(6) Modern life styles and their health aftermaths.

(7) Resources: qualitative, quantitative and indigenous perspectives.

(8) Environment, health, population dynamics and ecological imbalance.

(9) Inter- and intra-regional differentials in health spectrum.

(10) Health research centers and pharmaceutical interaction.

(11) Malaria, Cholera and diarrheal diseases: Incredible new microbes.

(12) Cancer and cardiovascular diseases: Occupational and environmental
impacts.

(13) Influenza, infectious diseases and HIV.

(14) Addiction to alcohol, narcotics and (prescription, OTC and illicit)
drugs.

(15) Quality of life, occupational health and outlook on life.

(16) Diabetes and other chronic diseases and disorders.

(17) Health awareness, attention and education.

(18) Oral hygiene, malnutrition, dental care and physical exercise.

(19) Obesity, arthritis, gout, autism, epilepsy and depression.

(20) Dementia, Alzheimer and Parkinson's disease.

(21) Rationality and feasibility of an universal plan.

(22) Epidemiology, demography and human development.

(23) Elderly, especially single people's plights.

(24) Health manageability and governance perspectives.

(25) Whither Statistical planning, rationality and interpretation.

Instead of an exhaustive listing of such items, it is statistically better to
sort out them according to their relative importance, choose a small number
of (how many?) most relevant ones, minimize the loss of information due to
such a subset selection, check its sustainability, and develop a manageable
and amendable operational plan.

3. Whither Statistics?

There is a need to incorporate statistical reasoning to chart out clinical and socio-economic diversity and resources) along with affordability and sustainability consideration. Fraudulent use of superficial universal health care plans and lack of affordability can create impasses. Lack of support from employer, government and other sources may be a stumbling road-block. All the items listed in Section 2 have serious impact in order not to affect abruptly or eliminate any major sector of the population. Borrowing health plans from the occidental countries may not be suitable for the orient. The plights of poor and rich are different across the global spectrum. Affluents have serious health problems too, while poors, particularly in the orient, in spite of having a sober outlook on life, may virtually wipe out the social equilibrium if the poor-rich inequality explodes. Most of the conventional economic measures and interpretations, currently in use, need to be critically statistically appraised.

China and, to a certain extent, India have contrasting features of spectacular economic growth in relation to human health and human rights: The industrial evolution in China is escalating its super-pollution status and its urban-rural economic disparity. In both the countries, only a fraction of over a billion population have benefited from this economic tide; more than 90 percent remain mired in poverty, thus leading to distributional disparity and environmental health concerns and deadlocks. Although, there is a gulf of differences between the two countries, can China avoid some of the problems post-independent India has had with labor(unions)? In the midst of current economic turmoils, how the health care system is shaping up in either country? India has some of the worst child mortality and maternal death rates outside the sub-Saharan Africa. This disastrous feature has invaded the *middle income class* (MIC), even the affluents. Public hospitals are running in inadequate and improper environment.The retired and elderly people are in desperate need for health care facilities. How any health care system can handle this delicate problem? Can the poor people afford to have any effective medical treatment within their means? A part of the MIC society is shrinking *below the poverty line* (BPL), no matter how we define and interpret the BPL, and the complementary part aspires to be in the *above affluence line* (AAL) at any cost to their social tradition and integrity. What universal health plans mean in such a sandwich socio-economic set-up? The interface of modern India or China has a glamoring side in information- and nano-technology, while the rest of socio-economically untouchables, inadmissibles and unreachables pose greatest threat to any social welfare and health care. Should we leave out specific sectors of a society in adopting a health plan? Can we ignore the possibility of a drastic breakdown of social system if a greater part of the society is left out? A statistical appraisal of this complex is a prerequisite for planning any health care system. It took considerable efforts on the part of governments in the West (Europe and North America) to formulate some social security or social welfare plans to include all sectors of the society and to provide Elderly medicare benefits, albeit most of them are far from being ideal.

4. BPL, AAL, MIC and Nested Inequalities

Inequality in diverse modules has persisted in all societies, at all times. Under such man-made multifactor complex, inequality may become so overwhelming that it can impede the progress or even the existence of a rational society. The compartmentalizing of any society based on such network of inequalities may invariably result in differential life style. Therefore, any development of health plans must take into account this intrinsic heterogeneity of human wealth, health and life. In a monetary interpretation, the diversity has led to poor, middle class, and rich people in terms of standard of living, and this in turn, started revolving around some manageable interpretation of wealth or real income. The classification of poor resulted in a fixation of BPL. Similarly, the interpretation of AAL, rests on an *affluence line*,leaving in between the two lines MIC, which may not feel comfortable to be classified as poor and at the same time may not afford to match the life style of the rich or affluent people. Further, the middle income class bifurcated into upper and lower middle class, in their differential objectives of life-style or standard of living. Even so, the qualm remains in tact: How to combat the threats to the tailored life-style, inducing unpreventable, unpredictable health problems. Health plans should cover these small pockets of ethnicity and must be flexible enough to suit this population diversity.

Leaving aside the rationality of drawing a poverty line solely based on some sort of income distribution, there is a pertinent question: Can we treat all the poor (or the rich) people alike? They differ in their income as much as in other physical and mental traits. As such, such hidden factors need to be taken into account in prescribing a meaningful interpretation of the BPL. It is a bit surprising that in lieu of the per diem protein intake, a surrogate variable, namely, per diem calorie intake is sometimes used to determine the BPL. Can the proportion of the calorie due to protein intake, different for animal and vegetable sources, be assessed fairly by looking at the total calorie intake? Can we ignore the vast number of explanatory variables or covariables in this assessment? What could be a reasonable statistical equation in a comprehensive modeling of BPL? Are we not to integrate health awareness, outlook on life, level of education etc. in the formulation of BPL? How are we to differentiate between various sectors in interpreting the BPL? What about the role of poverty indexes? Assuming that it is feasible to redistribute the wealth to a certain extent from the affluent to the BPL class, will it be better to do that in a uniform way or to have a specific way to improve the poverty index without changing the number of poors? We may refer to Chatterjee and Sen[1] for a comprehensive account of these inequalities. The present socio-economic dynamics have been steadily pushing up the affluence index by tilting the affluent income distribution to the right calling for valid, interpretable and efficient statistical resolutions, based on socio-economic, clinical and environmental perspectives.

5. Clinical and Environmental Perspectives

In any comprehensive and composite health assessment and health care plan it is essential to appraise epi-health status, thus highlighting the clinical perspectives. The evolution of bio-medical (clinical) sciences has been mostly governed by the empirics and heuristics; only in the recent past, some attention has been paid to the stochastics. Medical science is not an exact science. Yet, can we deny the vast amount of intuition and scientific outlook that reshaped medical science? Can we also deny that no two human being having similar symptoms of a medical problem would be identical with respect to the nature and degree of the problem, nor would require exactly the same treatment? It is this immense variability of human beings with respect to their metabolism, response to drugs and treatments, as well as, exposures to various health hazards that makes medical sciences more experimental in flavor, less precise than controlled laboratory experiments, and more difficult to administer for therapeutic and diagnostic uses.

At this juncture of time, medical science has annexed a wider interdisciplinary field of clinical sciences, biotechnology and *information technology* (IT). Yet,in spite of all our advancements in science and technology, major challenges have erupted from improper and careless treatment of our *bio-environment*, impacting *public health*, especially in the developing countries where resources-inequality may create roadblocks. Our bio-environment constitutes the entities of all socio-economic, cultural-political, clinical, biomedical, ecological and environmental (health and hazards) perspectives that are relevant to the dynamics of all bio-systems on earth, including mankind. The concept of *Quality of Life* (QOL), having its genesis in medical assessment of the status of patients undergoing some treatment for a chronic disease or disorder, has emerged as an essential tool for the assessment of the diverse impact of environmental and ecological hazards. Our mother planet is endangered with life-threatening phenomena, not only due to escalating ecological imbalances and environmental disasters, but also due to mounting social, economic, religious, geo-political and cultural disruptions. While we are combating with existing major health threats (like cancer, tuberculosis, gastro-intestinal diseases, cardiovascular and coronary blockage and strokes), new or hitherto unknown forms of catastrophic diseases or disorders (such as the HIV/AIDS) have invaded our life and posed enormous risk to our health and survival. Bioenvironmental toxicity of various kinds has attained an alarming level. Even mental health aspects deserve a close scrutiny in the light of bioenvironmental impacts.

Assessment of health-hazards from bioenvironmental toxic agents with a view to implementing (physical as well as mental) health promotion and disease prevention has been one of the major tasks of the public health discipline. However, realizing the far greater impact of our fast-deteriorating bio-environment, public health has led an interdisciplinary task where *environmental health sciences* and epidemiology, maternal and child health, nutrition, among others have combined forces towards a better understanding of the underlying complexities and better assessment of the risk. In the West, government regulatory agencies as well as non-government pharmaceutical research groups have also joined this task-force. The situation may be quite different in developing countries; even data-collection in a very objective fashion could be a problem. (Bio-)statistical reasoning is essential in this respect. Statistical *planning, modeling* and *analysis*, viewed from a broader public health and bioenvironmental health, play a fundamental role in this context. Hereditary factors in the orient are different from the West, so are the climatic factors. Empirics and heuristics should be an integral part in our drug-discovery and clinical practice. Jumping on the occidental train for a better ride to the orient could be bumpy.

Environmental toxicity (ET), emerging mainly through environmental pollution and contamination, has posed serious threats to the safe and prosperous propagation of mankind. The picture is highly complex due to multiple interacting sources of pollution and contamination, through a synergic combination of *absorption, inhalation* and *ingestion* toxicity. No country is immune to this disastrous xenobiotic effect. It may be argued that inhalation may be the primary industrial /occupational / environmental site of uptake of toxins. The water we drink may be contaminated with chemical dumping and arsenite compounds, the food we intake may have serious contaminations from pesticides, and above all, the air we inhale may contain significant amount of airborne particulate matters (APM), chemical (industrial) wastes and exhausts, automobile exhausts, gaseous emissions from garbage disposals and land-fill sites, environmental smoking effects, local or environmental tobacco (active and passive) smoking effects, thinning of the ozone layer (greenhouse effect), and an innumerable number of other factors, likely to cause toxicity of certain types. The intensity of ET has gone up dramatically in the past two or three decades, and awareness of this serious risk is a must for our survival. In fact, assessment of this risk with a view to minimize it to the extent possible is by far one of our most challenging scientific tasks. An interdisciplinary approach with due emphasis on statistical reasoning is essential for this endeavor (Sen and Margolin,[17] Sen[12-14]). Basically the different modes of toxicity intakes by human being are interrelated, and there is a need to have a more comprehensive statistical modeling that takes care of their synergisms as well as specificity to a desirable extent.

Quality of Air - Inhalation and Absorption Toxicology: Although some plants live on air, sunlight and moisture, most of the animals survive on all the three vital resources, namely, air, water and some form of food intake. Human respiratory system inhales oxygen-rich air, and at the alveoli inhaled oxygen is passed on to the minute blood vessels while carbon dioxide and water vapors are exhaled out through the mouth. As and when the inhaled air contains the APM some of which are toxic, there may be a gradual blockage of the exchange of oxygen and impure gas at the alveoli, resulting in toxicity at the blood level. Tobacco and other narcotic elements can trigger-off this toxicity process, and is supposed to be a principal conveyer of cancer and other lung diseases. Reduced oxygen to the cortex area also induces some malfunctioning of the brain as well as the central nervous system, and as a result, physical as well as mental health problems may crop up. Automobile and industrial exhausts add more undesirable elements to the surrounding air and they also lead to various diseases and disorders. In a different mode, our modern life-styles, increasing dependence on chemical and aerosol products, deforestation and uncontrolled energy consumption have all contributing to the thinning of the ozone layer; in turn, we are having the so called greenhouse effect. The increased ultra-violet level in the air through sun-light is a significant cause of absorption toxicology, and skin cancer as well as other epidermal diseases are associated with this effect. The elimination of lead from gasoline products has resulted in some improvement in the quality of air, though there remains much to ponder into the xenobiotic effects of ET. Burning of oil facilities in Kuwait twenty years ago and the Iraq-Afghanistan episode during the past 11 years resulted in an enormous ecological disaster, and some of the effects are still perceptible in affected population. In the Vietnam War, use of chemical warfare had similar impact. Emissions of radioactive fumes from Nuclear power plants as well as chemical toxic fumes from other industrial plants (Bhopal, India 1984) have a catastrophic impact and air was the main conveyer. Through the efforts of public health and environmental health scientists,it has come to the attention of governments, regulatory agencies, and public as a whole that such toxicities have also grave impact on human behavior as well as fecundity; the branch of *neurotoxicity* and *reproductive toxicology* deal respectively with these aspects. All of these have dominant genotoxic undercurrents, and the modern developments in *molecular biology* and *genetics* are casting light on such perspectives. And yet, not much is known about this complex ET phenomenon.

Food, Drink and Drug Ingestion Toxicity: Massive use of fertilizers and pesticides has increased the food grain production and storage preservation capabilities, and yet, a significant percentage of agricultural products gets unusable due to poor distribution from the original sources to the markets, and more significantly at the cold-storage facilities. At the present, in the West, genetic technology is incorporated in achieving further revolution in food production and distribution. But, there are some concerns which are coming to the surface at an increasing rate. Increasing carcinogenic activities of the food preservatives and pest control products are being detected. The human consumption of red meat is raising question about the health effect of fat and elevation of blood-level cholesterol. The battery of cardiovascular and coronary diseases has found their link to consumption of high-fat high-cholesterol food items including the fried foods. In the drink sector, even the so called soft-drinks may have adverse health effects. Alcoholic drinks, apart from affecting the central nervous system and destroying thousands of cells in the cortex, induce intoxication that could indirectly lead to devastating effects. In the West, death and disability from automobile accidents on road is a significant cause of human casuality, and a majority of these are actively or passively due to drunk driving. Drugs other than prescription medicine are more notorious. Narcotic drugs and illicit drinks have been stamped as an alarming international threat. Even on the medicinal front, many of the prescribed drugs are known to have some (and often serious) side-effects. The use of contraceptives has become so widespread that it is unthinkable how for thousands of years human society survived without such devices. The escalating incidence rate for breast and uterus cancer for women can be accounted to a greater extent as linked to such practice. On the other hand, there is a natural need for estrogen intake for women at a certain stage of their life, and yet we can not rule out carcinogenic side-effects from such use. Faced with our task to control the growth of human population on earth, the basic dilemma is whether or not to go for population control? It's a strange bio-environment. During the past 55 years, we have literally surrendered to antibiotics; they used to work wonder against some diseases that were once thought to be incurable! But, after 50 years, we see that their potency is decreasing at an exponential rate, and microbes that are antibiotic-resistant are coming back at an alarming rate. We claimed some twenty years ago that due to indiscriminate use of DDT and other products, malaria has been eliminated from earth. But during the past few years Malaria is back more virulently than before. It is not out of the way to mention here about the HIV/AIDS problem that is so widely prevalent at the present time, and drug and abnormal sexual practice related that unless abated it could completely reshape our bio-environment with far reaching and deleterious effects.

The Water Crisis: Water is indispensable for human life on earth. For biological and physiological needs, we need to drink a lot of water, and we need to use water for cooking and sanitation purposes. Water is needed for our agricultural as well as industrial needs, and it is needed for any other task in life. And yet, in this desperate need, we have ignored the basic alarms : what quality of water can we expect with all our misuses? Basic water crisis problems are the following:

(i) Water contamination due to industrial as well as human waste dumping in the waterways (rivers, lakes and even seas).

(ii) Increasing deforestation resulting in more soil erosion not only at the origin of rivers but also in sediment deposition all along their way, having high level of bacterial and other arsenic contents.

(iii) Inadequate handling of recycling of water for urban needs (the sewage problem).

(iv) Natural disasters like flood, hurricane, tornado which sweep away all toxic (decomposed) ground substances to waterlines, and on top of that, quite often have dead animals and open sewage in the water streams contributing an enormous ecological problem.

(v) Industrial (subsoil) dumping that leads to highly toxic materials which find their way to water channels, and disrupts marine biology (killing of fishes etc).

(vi) Deep tube-well and arsenic subsoil material (rehabilitation areas), and

(vii) oil spills in coastal waterways that also affect marine biology, and thereby contribute to the water problem. The dry-docks in port-cities for painting the outer hull of boats and other maintenance facilities are dumping toxic and arsenic materials in the river mouth, and often carried for miles in the upstream inlets. The climatic changes all over the world (including the greenhouse effect) are also contributing to the deterioration of water quality. Increasing demand for water consumption from urban and metropolitan areas, even in technologically advanced countries, and subsoil ground arsenic contents of some areas also contribute to this health crisis.

Quantification and Assessment of QOL: It is a composite task to define quantitatively the QOL in the context of bioenvironmental impacts and public health concerns. Much of this assessment relates to a quantification of some qualitative states in a form that would permit statistical interpretation, modeling and analysis in a simple and yet valid manner. To a socio-economist, QOL would be more appropriately interpretable in terms of standard of living with natural emphasis on affluence, poverty and income inequalities. Nevertheless, the quantification of real income or wealth is itself a very delicate task, and converting the life-style and social environment of people from different walks of life into comparable income is often highly controversial, and the top economists are still struggling to tune it in a mutually agreeable norm. The damage to our bio-environment is mostly due to our addiction to modern life-style, not caring much about the environment. This has created HRQOL (health related quality of life) problems that governments and public health policy and administrative agencies are increasingly concerned with; we need to pay more attention to HRQOL.

6. Statistics: Perspectives and Task Ahead

In the assessment of the bioenvironmental state of earth and the imposing health risks, in addition to bioenvironmental factors that can be easily identified and quantitatively assessed properly, there are scores of less apparent or more qualitative ones; it needs to be settled how all or some of these are relevant. Keeping these in mind, we may categorize the main role of statistical risk assessments of bio-environmental hazards into five sectors:

1. Inventory of sources of toxicity and quantification of intensity of specific items.
2. Acquisition of scientific evidence of dose-response relations through laboratory (dosimetric) studies mostly involving subhuman primates.
3. Collection of information on human experience through epidemiologic (observational) studies.

4. Data collection, data-quality control and statistical monitoring of dosimetric or observational studies.

5. Pooling of statistical evidence from acquired datasets in dosimetric and observational studies (meta analysis).

The discipline of epidemiology has a great bearing in this respect. Epidemiology aims to combine information acquired from etiologic, toxicologic as well as observational studies. For this task, statistical reasoning is essential in the planning of such a study, in developing statistical models that take into account epidemiologic undercurrents fully, and to draw statistical conclusions in an objective manner.

As in most of other clinical and biomedical studies, *risk analysis* can only be made from experimentation (e.g., clinical trials, laboratory experiments, tissue culture etc.,) or observational data set , in either case, under well defined objective plans that permit valid and efficient statistical conclusions. In this respect there are various points of difference between laboratory experiments and bioenvironmental studies, the latter being under much less controlled experimental or observational setup. In this perspective, the design of bioenvironmental studies may be quite different from conventional agricultural or biometric studies, and the associated sampling schemes could be quite complex. In fact, the units in a bioenvironmental study may have a number of auxiliary or explanatory variables in addition to the primary response variable(s), and they usually have profound effects on the response pattern. Therefore, in statistical modeling, these explanatory variables need to be incorporated in a suitable way; this introduces complications in the design and modeling of bioenvironmental studies. Secondly, in standard statistical models, it is often assumed that apart from the systematic component the response variables have chance components or errors that are independent and identically distributed. This stringent assumption is most likely to be untenable in bioenvironmental studies. As a matter of fact the very process in which data are collected can induce considerable complications in the sampling design and associate statistical models. For example, in air pollution studies, suppose one wants to record the level of carbon monoxide and other pollutants. This might depend very much on the proximity of the site to emitting sources; it could be also different at different time-points and different altitude from the ground level. Further, the question of practicality of recording such observations remains open. It's quite different from measuring rainfall at a place! Therefore, the intricate structure of sampling design has a great bearing on statistical modeling, and monitoring of data quality is an essential task. There is also a visible spatio-temporal variation structure in most bioenvironmental studies. In conventional studies, it is generally assumed that there is a homogeneity with respect to spatial dependence, and often, the *kriging* methodology involving variogram models yields manageable statistical resolutions. However, in the absence of such a spatial homogeneity, the kriging method may lose its appeal to a greater extent. Moreover, the conventional linear models may not be tenable in bioenvironmental studies (as often the response variables are count variables with distinctly skew distribution and nonlinear systematic components. For this reason, in bioenvironmental studies, sometimes, suitable *generalized linear models* (GLM) are used. However, in the presence of various auxiliary variables, such GLM could be very complex and not practical to a certain extent.

There is a greater problem with statistical modeling and analysis of bioenvironmental studies. With the able help from environmental scientists, in the West, it has the machinery of having a network of stations or sites where environmental pollution data are recorded on a regular basis, and referred to the regulatory agencies for their monitoring of the environmental pollution. On the other hand, there are also various epidemiologic studies, mostly observational, that aim to relate mortality and morbidity pictures to the environmental pollution and other hazard picture, and thereby to prescribe the impact of bioenvironmental hazards on human health and life. This picture can not be complete without the missing link: how specific environmental factors are related to such epidemiologic findings. For example, there is a common belief that smoking (active or passive) is related to lung cancer. We may gather the prevalence of smoking in an area and relate it to the incidence rate of lung cancer in the same area. But, without an etiological information (or causal-effect), how can we relate high cancer rate to higher proportion of smokers? There could be some other explanatory variable(s) (like familial effect and drinking habits) which might provide a better explanation. Of course, it is often difficult to conduct an experiment involving human subjects for drawing statistical conclusions. For that reason, first some laboratory studies are made on subhuman primates, and their findings are then extrapolated to human beings. This comes under the jurisdiction of dosimetry and animal studies. Even so, in terms of metabolism and other factors, human response could be quite different from such subhuman primates, and hence the statistical conclusions can not be transmitted to human being without critical appraisal of physiological as well as environmental differences. Towards this goal, during the past thirty years or so, in the West, controlled clinical trials have been administered on selected human groups, and guided by their findings more thorough appraisal of biological factors are planned. In terms of statistical modeling and analysis, such dosimetric studies and clinical trials have revealed some methodological challenges, and some resolutions have been made. In this respect, traditional parametric models have been deemphasized and nonparametric as well as semiparametric methods have been advocated. A good deal of such findings can be found in the various chapters in the volume edited by Sen and Rao.[18]

The basic methodological issues relating to the health hazards from environment and ecosystem, have led to use and abuse of statistical packages which have literally flooded the users' market. Basically, information technology has led to mechanisms to collect and analyze huge data sets. However, as has been pointed out before, the process of data collection is itself very important for statistical modeling, and on top of that statistical analysis should conform closely the underlying methodology. Unfortunately, because of model complexities and nonstandard methodologies, in many cases, use of standard statistical software packages could be misleading and lead to imprecise or even inconsistent conclusions. Therefore, we need to develop side by side appropriate statistical software packages that could be used more validly in bioenvironmental risk analysis.

7. Integrating Measures of Socio-economic Features

Currently adopted measures of socio-economic inequality, and poverty indexes take into account the proportion of population belonging to such a class, their mean or median level and some inequality measures, such as the Gini coefficient. Basically, there is a subtle quantification of the underlying response variable rendering such measures to be well defined. In a broader context of diversity or distributional inequality, there may be relatively more prominent qualitative flavor which, in turn, raises the question on the suitability of moment-like measures in such formulations. An extreme case is a multi-dimensional contingency table, without an inherent quantitative trait, exhibiting only qualitative variation. On the other hand, in poverty indices there is a subtle monotonicity condition (under (*stochastic ordering*) and violation of which may diminish the rationality and utility of an index. Chatterjee and Sen[1] proposed some modifications and developed some stochastic ordering monotonicity preserving diversity measures which are suitable even for qualitative data having some partial ordering properties. We examine some *Lorenz ordering* and stochastic ordering properties in this broader context.

Socio-economic Features: We live in a world that is intermittently going through changes in its value system, ethics, religious faiths, cultural outlooks, and above all evolutionary socio-economic (social welfare and economic policy) deformations. Yet in this environment we speak of economic well being, social equity and equal opportunity - though they could be highly illusive. In resources inventory, developmental economy, planning and in other problems too, we have a strong mathematical economics flavor; it is of interest to appraise the role of stochastics in such deterministic models. On a regional or country basis, the governments have the responsibility to fathom out the state of economic and social well-being, and in this respect, even health-economics is not outside the pandora's box. On top of that international agencies like the United Nations, UNESCO, FAO, ILO, IMF, and World Bank, aspire to have this quantitative assessments across the nation or regions, and over the passage of time as well. The fancy term QOL, albeit having its genesis in health management and care, is being used in a much broader field of socio-economic investigations. The game is to quantify all these more or less qualitative features, and socio-economists take great pride in this task, yet the key role of statistical science can not be ignored. There are some compatibility issues, and some other intricate features that definitely merit critical statistical appraisals.

Let us concentrate first on wealth, real income and the assessment of poverty as well as affluence of a community or society in the midst of a mixture of the so called poor people, the rich or affluent ones, and the go-in-between middle class people. The first and foremost task is the assessment of real income or wealth. In this assessment there may be quite a number of explanatory or auxiliary variables, many of which could be qualitative (or at best polychotomous categorical). Therefore, there is a genuine need to formulate a single criterion variable that relates to the real income or wealth on a personal or household basis; in the latter case, the size of the household is an important explanatory criterion too.

Let us denote the real income variable by X, and assume that X has a distribution function (d.f.) F, defined on $\mathbf{R}^+ = (0, \infty)$. Also, with respect to this d.f. and with due consideration on the real interpretation of poverty and affluence, we conceive of two numbers $(0 <)\rho < \xi(< \infty)$, termed the *poverty line* and *affluence line* respectively, such that a person (household) having a real income below ρ is termed poor, and above ξ is termed rich. With that in mind, let us define

$$\alpha = F(\rho) \text{ and } \gamma = 1 - F(\xi), \tag{1}$$

as the proportion of poor and affluent people respectively. In the simplest way, these numbers are often used to define the well-being of a society or community or country. However, they do not reflect the extent to which poverty and affluence persist. To have a more meaningful treatise, we therefore need to look into the poor (and affluent) income distributions. We consider here the case of poverty, and a very similar case can be considered for affluence. Let

$$F_P(x) = \alpha^{-1} F(x), \ x \leq \rho, \text{ and } 1, x \geq \rho. \tag{2}$$

As in Sen,[11] we consider the rescaled poor income distribution

$$F_P^o(y) = F_P(y\rho), \ 0 \leq y \leq 1. \tag{3}$$

Further, let

$$\mu^o = \int_0^1 y dF_P^o(y), \ G^o = (2\mu^o)^{-1} \int_0^1 \int_0^1 |y - z| dF_P^o(y) dF_P^o(z) \tag{4}$$

be the mean and Gini coefficient of the d.f. F_P^o. Then a commonly used poverty index, due to Sen,[7] is the following:

$$\pi_S = \alpha\{1 - \mu^o(1 - G^o)\} = \alpha\{G^o + (1 - G^o)(1 - \mu^o)\}. \tag{5}$$

Since μ^o, G^o both lie in the unit interval $(0, 1)$, π_S lies in the interval $(0, \alpha)$, and it reflects the impact of both the mean income and income inequality among the poor. A larger G^o for a given μ^o makes the index larger, while for a given G^o, π_S decreases with the increase of μ^o. Of course, the distribution being defined on $[0, 1]$, the parameters μ_o and G_o are interrelated and their covariation plays a basic role in the interpretation of the poverty index. From robustness perspectives, Sen[9] considered another version wherein the arithmetic mean is replaced by a geometric mean:

$$\pi_S^* = \alpha\{(1 - \mu^o)^{1-G^o}\}. \tag{6}$$

Note that π_S, π_S^* are unit-free (as is F_P^o), but not necessarily invariant under arbitrary increasing transformation on the original income variable. This last property is important especially when income has a significant qualitative component that may make it difficult to quantify it precisely and hence provokes ordinal categorical data models (see Sen[11]). It is not uncommon to have the poor income distribution on a set of ordered class intervals, and bearing in mind the qualitative undercurrents, a categorical data model seems to be quite reasonable.

With that motivation, Sen[11] pursued the idea of incorporating the Gini-Simpson diversity index in a more general utility-oriented formulation and proposed some allied measures. These are also related to the so called *diversity measures* considered by Rao.[6] Nayak[4] and Nayak and Gastwirth[5] used Rao's quadratic entropy measure in a distributional (inequality) setup, but confined to ordered categorical data models. Consider a categorical data model relating to $C(\geq 2)$ categories, indexed as $1, \ldots, C$ with respective probabilities $P_1, \ldots, P_C$, so that the point $\mathbf{P} = (P_1, \ldots, P_C)'$ belongs to the C-simplex $\mathcal{S}_C = \{\mathbf{x} : \mathbf{x} \geq \mathbf{0}, \mathbf{x}'\mathbf{1} = 1\}$. The Gini-Simpson index (of diversity) is defined as

$$I_{GS}(\mathbf{P}) = 1 - \mathbf{P}'\mathbf{P} = \sum_{j=1}^{C} P_j(1 - P_j) = \sum_{1 \leq j \neq k \leq C} P_j P_k. \qquad (7)$$

A natural extension (see Sen[11]) of this is the utility-oriented one:

$$I_{UGS}(\mathbf{P}) = \sum_{j=1}^{C} u_j P_j(1 - P_j), \qquad (8)$$

where the u_j stand for some utility scores. I_{GS} is a special case of Rao's quadratic entropy measure:

$$I_R(\mathbf{P}, \mathbf{D}) = \sum_{j=1}^{C} \sum_{k=1}^{C} d_{jk} P_j P_k, \qquad (9)$$

where the d_{jk} stand for the distance between the categories j and k. Thus, the d_{jk} are all nonnegative, and $d_{jj} = 0, \forall j$; if we set $d_{jk} = 1, \forall j \neq k$ then $I_R = I_{GS}$.

A poverty or income inequality index serves as a summary measure of the income distribution (of the poor). As such, it is generally used in comparative studies of poverty or income inequality of different regions or different communities as well as for temporal variation of the same in a given setup, or more generally, in a spatio-temporal study covering a greater area and a time period. In order to satisfy the basic compatibility criteria, there are certain properties (or axioms) that such measures should possess. Among these axioms, the two most important ones are the following:

(A) An index should be nondecreasing under left-shift of the reduced income distribution (F_P^o), or in other words, it should be nonincreasing under a stochastic ordering on F_P^o.

(B) An index should be subgroup or ANOVA decomposable, that is, if a population is composed of several subpopulations, then the related overall index should be decomposable into two nonnegative components representing respectively the average of the within subgroup measures and the between subgroup divergence.

The last property has been extensively studied in the literature (see Nayak,[4] Nayak and Gastwirth,[5] Foster and Shorrock,[3] Foster et al.,[2] Sen[8] and Sen[11]). All the measures referred to earlier satisfy this subgroup decomposability condition. Shorrocks[19] considered some additively decomposable inequality measures; there is a persistent quantification requirement for which additivity makes sense and entropy functions play a basic role in this context. Coming back to Rao's quadratic entropy measures, we may represent this in the form

$$Q(F) = \int \int d(x,y)dF(x)dF(y), \qquad (10)$$

and in the present context, we work with the rescaled d.f., so that the range of integration is $(0,1)$. Suppose now that a rescaled d.f. F is a mixture of two such d.f.'s, say F_1, F_2 with a mixing coefficient $\omega : 0 < \omega < 1$. Then writing $F = \omega F_1 + (1 - \omega)F_2$ we have

$$Q(F) = \omega Q(F_1) + (1 - \omega)Q(F_2)$$
$$+ \omega(1 - \omega) \int \int d(x,y)d[F_1(x) - F_2(x)]d[F_2(y) - F_1(y)] \quad (11)$$

so that in order that the decomposability holds (where the first and second terms on the right hand side represent the within and between components respectively), we need that

$$\int_0^1 \int_0^1 d(x,y)d[F_1(x) - F_2(x)]d[F_1(y) - F_2(y)] \le 0. \qquad (12)$$

This is known as the *conditionally negative definiteness* (CND) condition. Equivalently, we can characterize the CND condition in terms of $Q(F)$ being a concave function on the space of distribution functions. It is also possible to characterize it in terms of $d^{1/2}(x,y)$ being a metric. Whereas this CND condition is easy to verify, it may not necessarily satisfy the monotonicity criterion in (A); the index I_{GS} is a classical example for this.

Let us examine the monotonicity property in (A). Suppose that there are two rescaled d.f.'s F_1, F_2 which are stochastically ordered, *i.e.*,

$$F_1(y) \geq F_2(y), \ \forall \ y \ \in \ (0,1). \tag{13}$$

Let $q_j(y) = \inf\{x : F_j(x) \geq y\}, \ j = 1,2$ be the two *quantile functions*. Then, note that the stochastic ordering property is equivalent to

$$q_2(y) \geq q_1(y), \ \forall \ y \ \in \ (0,1). \tag{14}$$

It is easy to show that $\pi_S^{(j)} = \pi_S(F_j)$ can be written as

$$\pi_S^{(j)} = \alpha\{2 \int_0^1 \{1 - q_j(y)\}(1 - y)dy\}, \ j = 1,2, \tag{15}$$

when both the d.f.'s are continuous. As such, in this case, by the last two equations

$$\pi_S(1) \geq \pi_S(2) \text{ under } F_1 \geq F_2 \text{ a.e.} \tag{16}$$

The picture could be different when the d.f.'s are not continuous (as is typically the case in the current context).

It is also possible to compare the d.f.'s F_1, F_2 directly without comparing them by a single measure (such as the π_S). We define the mean μ and quantile function $q(y)$ as in above and denote by

$$L_p = \int_0^p xdF(x)/\mu, \ p \in (0,1). \tag{17}$$

The graph $(p, L_p), \ 0 \leq p \leq 1$, is known as the *Lorenz curve*; it lies below the diagonal line $L_p = p, 0 \leq p \leq 1$, touching the two tips $(0,0)$ and $(1,1)$, and the area formed by the Lorenz curve and this diagonal line is termed the concentration index. If there are two rescaled d.f.'s F_1, F_2, we say that a Lorenz ordering holds if one curve lies above the other; in symbols

$$F_1 \succ_L F_2 \text{ if } L_p^{(1)} \geq L_p^{(2)}, \forall p, \tag{18}$$

where $L_p^{(j)}$ refers to the Lorenz curve for $F_j, \ j = 1,2$. There has been an extensive statistical literature on Lorenz ordering (Foster and Shorrocks[3]). In the present context we may like to know whether stochastic ordering and Lorenz ordering are isomorphic, and if not whether or not one implies the other. The answer is in the negative (Chatterjee and Sen[1]).

In all these measures, there is a natural emphasis on the actual income variable (rescaled to the poverty level) or suitable quantitative utility scores.

There are some situations without a precise quantification, resulting in pure qualitative categorical data models. These refinements are considered below.

Health Economics and Qualitative Poverty: In 1993, WHO has launched a Project WHOQOL on QOL to have a global picture of human health and the impact of the bio-environment on it. QOL has been interpreted mostly in a health related perspective, though this interpretation is undergoing an evolution in a much broader perspective. In the WHO interpretation QOL is defined as an individual's perception of their position in life in the context of the culture and value system in which they live and in relation to their goals, expectations, standard and concerns. The QOL interpretation has also been localized for a more specific subclass of population. Consider, for example, a class of people in a specific geo-political area who have undergone a major medical/surgical treatment (e.g., cancer, heart problems, chronic disorder, depression, and many other long-effect diseases/disorders), so that for the rest of their life, their physical, mental and functional abilities may be impaired to a greater or lesser extent. Parallel to the case of economic poverty, the main concern here is the quality of life (i.e., the deficiency) these people have at this stage.

As in the case of poverty, a crude measure such as the proportion of the population afflicted with such a disease or disorder does not convey a representative picture of the actual state of affair with these people. A second possibility is to incorporate the measure MRL (*mean residual life*) that attempts to compare the normal cohorts with the afflicted ones. However, that may still be not totally representative. For example, for a common disease/disorder there may be two treatments, say A and B. Treatment A is known to be effective in prolonging the MRL albeit without much care of the actual functional ability of the afflicted persons, while Treatment B is primarily geared to enhancing the functional ability though possibly compromising on the MRL. For this reason, *quality adjusted life* (QAL) has emerged as a more usable measure that takes into account the longevity following such a disease/disorder as well as the state of their health status. In most of these cases, we typically have a multidimensional categorical data model with some (partial) ordering of the categories, and the basic statistical issue is to take into account both the QAL and MRL with a view to formulating a more meaningful measure (see Sen[10,15,16]).

Chatterjee and Sen[1] have considered a general class of indices with special emphasis on categorical data models without possibly a linear ordering of the categories. We can express such an index as

$$I_{CS} = \alpha\{\sum_{j=1}^{C}\sum_{k=1}^{C} d_{ij}P_iP_j\} = \alpha\{\mathbf{P'DP}\}, \tag{19}$$

where the C categories have respective probabilities $P_1,\ldots,P_C$, $\mathbf{P} = (P_1,\ldots,P_C)'$ and $\mathbf{D} = ((d_{ij}))$ is a symmetric matrix with nonnegative elements; α stands for the proportion of the population that is under consideration (e.g., poor people in poverty studies). The poverty index π_S is a bonafide member of this class where the matrix $\mathbf{D}$ has a definite quantitative structure that depends on the individual cell variate values. Chatterjee and Sen[1] have shown that under an *increasing north-westerly trend* (INWT) property of $\mathbf{D}$, the stochastic ordering property of π_S extends to this entire class.

Let us turn back to the QOL context where there be a more dominant qualitative flavor for the different categories, though it might be possible to identify an underlying partial ordering. This provocates the use of suitable grades or utility scores for the categories and replacing the role of income-gap ratio by such grade-gap ratio, analogous indexes can be formulated. The adaptability or appropriateness of such an index would depend on the appropriateness of the chosen utility scores. When the categories are only partially ordered, we use the notation $i \prec j$ to denote that category i is worse-off than j (with respect to the trait under consideration). Then, a special index advocated by Chatterjee and Sen[1] is the following:

$$I_{CS}^o = \alpha\{1 - \frac{1}{C}\sum_{i=1}^{C} (\sum_{j:i\prec j} P_j)^2\}. \tag{20}$$

This index avoids the use of arbitrary scores and incorporates solely the underlying partial ordering of the categories. In the context of *QOL-deficiency* (QOLD) such a measure can therefore be advocated. For example, for people identified in the HIV positive class, there are various categories depending on their physical and mental states with the worst state being the death-bed, and the best is the afflicted but still physically fit to meet daily living task. In such a case, the partial ordering is well identified, and hence I_{CS}^o can be used to have an interpretable QOLD measure. In a broader context, if only some partial ordering is preserved it may be better to conceive of some utility scores and define an index in the above fashion. In health measurements, as has been discussed earlier, the complexity of categories and their sheer dimension may dictate the use of Hamming distance type composite measures but possibly using utility scores in the different coordinate variables or attributes. Cost effectiveness and absorbability considerations may also dictate the use of suitable utility scores. More composite measures may be needed to suit the purpose better. Whereas in the case of income distribution, the classification of poor and others can be made conveniently, in QOL and human development assessments such a clear cut ordering is lacking and more socio-economic as well as statistical considerations are genuinely needed to formulate a rational classification of poor to reasonable norms. Sans this acceptable norm of classification, any routine use of crude measures could be highly unreliable to imprecise.

Developmental Economics and Measures : Developmental economics include a wide area where natural resources as well as socio-economic and geo-political factors all show up in a rather complex synergy. The associated explanatory as well as response variables may not be all quantitative and there may be a persistent qualitative component. In the context of national health plans, we can not disregard a vital component : mental health, outlook on life and social compatibility. At the present time, the family structure and family life are undergoing drastic change. The acute competition starting from the kindergarten schools all the way to secondary and college level education (or better training!) has created an enormous stress at all levels; not only the pupils in this phase of life are going to frustrations of all kind but their parent(s), often single, are going through an ordeal of mental tests and social incompatibility. This has resulted not only a disarray of attitude towards life but also a tremendous unrest, especially among the younger people. With most of the educated people, both the spouses are now a days working outside the home and their ability to adjust with the day to day family life is also put to test. Infertility among the working couple, more in the upper middle income group, has become a significant factor of modern family life. Our toxic environment is also contributing towards reproductive toxicology. Developmental biology is receiving good attention from researchers from all walks of social and clinical sciences. Human development is possibly at crossroads with this stress and strain in our daily life.

Mental health problems are cropping up at an escalating rate. Terrorism, be it at the domestic or geo-political level, has become a household word, albeit we may not be completely aware of the depth of this problem and its deleterious impact on our society. Modern life-style has also contributed towards this mental health problem and occasional outburst of violence thereof. Congenital mental health problems are also noticeable to a greater extent now than fifty years ago. The transformation from resource sharing joint families to small family units has created major social impasses. Even the parents of working couples are often left out in isolation with their plights at their golden age placed at the disposal of their destiny. Ironically, many of these parents sacrificed much of their own comforts to raise their child(ren) beyond their means for full education and did not realize that at their golden age, if they survive, they will be left-out! Dementia and Alzheimer disorder are not rare in Indian societies, and by no means, they are confined to poor people. Ironically, even among the offsprings, autism is perceived more often now than in the past. Is it really due to the changing environment or the life-style of present day youths? Can these problems be characterized totally by some quantitative disorder or disability? It has been perceived in the West that various irregularities of life-style at the pregnancy stage (of mothers) may affect significantly the chance of autism or other birth-defects and disorders. Can we borrow the basic idea of monetary poverty and formulate parallel measures of mental poverty? To what extent this is linked to human development? Even in the most simple formulation of this mental health phobia, though in principle, the situation is comparable to the poverty assessment task, because of qualitative factors, there is a genuine need to address them in a somewhat different manner. In this respect, the measures discussed earlier provide a convenient channel for formulation of suitable indexes that have certain properties analogous to the QOLD and QAL measures.

Alas, even in that simple setup, the assessment of mental health problems is a highly technical task which can not be handled without the formation of mental health plan provisions. Family members, due to social customs, may be shy in reporting any such disorder at the beginning stage, and at a later stage, it may be difficult to put a tab on this disorder. In the West, there are provisions for such autistic children to have school education under supervised guidance, followed by suitable job opening when they are ready to enter that phase. A national health plan must have sound provisions for such people, nurturing the development of mental facilities of these people and formulate suitable solutions so that they feel comfortable in being a bonafide part of our society. Along with psychiatric treatment, statistical considerations are of utmost importance in this challenging task.

Coming back to the developmental biology, I could make a comment that often the slow progression of human development is linked to extraneous environmental factors - not necessarily due to congenital causes. However, the association of genetic factors with environmental ones can make the phenomenon much more severe and unmanageable. This is the reason why in developmental biology it is important to sort out the genetic basis clinically as far as possible and then create a compatible environment for the nurture of their full human development. At the present time, realizing the steady growth of this disorder, health economics, clinical sciences and psychometric as well as statistical analysis should be harnessed together, to the extent of permissible resources. Emotion and intelligence quotients are not one-to-one related but there is ample room to accommodate them in a healthy environment. Society can not run on either emotion or machine intelligence. Health statistics and economics should harness emotion and intelligence together and heed to this mental health fortification as well.

A greater task is to take into account the different statistical aspects and combine them in to a composite issue to tackle the complexities of resources, need and sustainability of health care. Statistical allocation of provision for various medical needs and their level of coverage is indeed a complex task. Sustainability and affordability are also complex statistical issues. It is our desire to go for more sophisticated statistical appraisal of health plans and health care in regional and global setups in future communications.

Acknowledgments

The author is grateful to Professor Y. P. Chaubey for his careful editing and helpful suggestions.

References

1. S. K. Chatterjee and P. K. Sen, *Calcutta Statist. Assoc. Bull.* **50**, 137 (2000).
2. J. Foster, J. Greer and E. Thorbecke, in *Measurement of Inequality and Poverty*, Ed.: S. Subramanium (Oxford University Press, Delhi, 1997).
3. J. Foster and A. F. Shorrocks, *Euro. Econ. Rev.* **32**, 654 (1988).
4. T. K. Nayak, *Sankhyā* **B48**, 1 (1986).
5. T. K. Nayak and J. L. Gastwirth, *J. Bus. Econ. Statist.* **7**, 453 (1989).
6. C. R. Rao, *Sankhyā* **A44**, 1 (1982).
7. A. Sen, *Econometrica* **44**, 219 (1976).
8. A. Sen, *On Economic Inequality*, extended edition (Clarendon Press, Oxford, 1997).
9. P. K. Sen, *Journal of the American Statistical Association* **81**, 1050 (1986).
10. P. K. Sen, *Journal of Applied Statistical Sciences* **4**, 232 (1996).
11. P. K. Sen, *Calcutta Statist. Assoc. Bull.* **49**, 1 (1999).
12. P. K. Sen, in *Encyclopedia of Environmetrics*, Eds.: A. H. El-Shaarawi and W. Piegorsch (John Wiley, London, 2001), 1-4.
13. P. K. Sen, in *Encyclopedia of Environmetrics*, Eds.: A. H. El-Shaarawi and W. Piegorsch (John Wiley, London, 2001), 1054-1064.
14. P. K. Sen, *Current Sciences* **80**, 1167 (2001).

15. P. K. Sen, in *Proceedings of the Bernoulli-ISI Environmental Conference, Calcutta*, Eds.: A. K. Ghosh and J. K. Ghosh (Oxford University Press, New Delhi, India 2002).

16. P. K. Sen, in *Statistical Methods for Quality of Life Studies: Design, Models and Analysis*, Eds.: M. Mesbah *et al.* (Kluwer, Amsterdam, 2002).

17. P. K. Sen, and B. H. Margolin, *Sankhyā* **B57**, 252 (1995).

18. P. K. Sen, and C. R. Rao (Eds.) *Handbook of Statistics, Vol. 18: Bioenvironmental and Public Health Statistics* (Elsevier, Amsterdam, 2000).

19. A. F. Shorrocks, *Econometrica* **48**, 613 (1980).

A UNIFIED APPROACH BETWEEN POPULATION-AVERAGED AND CLUSTER-SPECIFIC ANALYSES FOR MULTILEVEL ZERO-INFLATED COUNT DATA

G. SNEDDON

Department of Mathematics and Computer Science, Mount Saint Vincent University,
Halifax NS B3M 2J6, Canada
E-mail: gary.sneddon@msvu.ca

M. T. HASAN and R. MA

Department of Mathematics and Statistics, University of New Brunswick, Fredericton,
NB, E3B 5A3, Canada

In medical and health research, excessive zeros often occur with multilevel longitudinal count outcomes. The zero and positive components of such longitudinal data are traditionally modeled by incorporating separate random effects into two part models such as zero-inflated Poisson or hurdle models. To accommodate zero and positive components in the multilevel longitudinal count response compatibly, we connect both components using a single random effects distribution, the compound Poisson distribution, in the zero-inflated Poisson mixed models. The compound Poisson distribution is semi-continuous with a positive probability mass at zero. Since there is also likely serial dependence in the longitudinal data besides over-dispersion modeled by the compound Poisson random effects, we incorporate additional random effects with an autoregressive correlation structure. Our modelling approach for zero-inflated multilevel longitudinal count data can unify population-averaged and cluster-specific analyses. A quasi-likelihood approach has been developed in the estimation of our model. We illustrate the methodology with analysis of the multilevel longitudinal hospital utilization data.

1. Introduction

Excessive zeros in longitudinal count data occur in variety of applications (see Hasan and Sneddon,[2] Lam *et al.*,[6] Min and Agresti,[13] and Ver Hoef and Jansen[18]). Although cross-sectional discrete data with excessive zeros have been studied extensively, it is only with recent work (for example, Hasan *et al.*[3] Lam *et al.*,[6] Ma *et al.*,[10] and Ver Hoef and Jansen[18]) that techniques for correlated data with excessive zeros have been studied. A zero-inflated Poisson (ZIP) model was first introduced by Lambert[7] to an application in manufacturing studies. Hall[1] extended the ZIP model by only introducing a random intercept to the Poisson mean to account for the within-subject clustering effects. Mullahy[15] introduced random effects into hurdle models, which can be used to analyze both zero-inflated and zero-deflated responses. A pair of uncorrelated random effects for both zero and truncated Poisson components of ZIP models was introduced by Yau and Lee.[19] Under the assumption that the zero and non-zero components for a subject would be correlated, Min and Agresti[13] used two separate but correlated random effects for these components. Min and Agresti[14] later incorporated such two parts random effect modelling approach in the hurdle models using a maximum likelihood approach. However, they noted that it could be challenging to implement with complex random effects structures. Lu *et al.*[9] pointed out that in these 'two parts' models most of the zeros are separated from the rest of the data as they are assumed to be structural.

This work was motivated by the health-care utilization data collected longitudinally from 1985 to 1990 by St. John's General Hospital in St. John's Newfoundland, Canada (Ma *et al.*[10]). In this data set, 48 families were observed for six years and the number of yearly visits to health care facilities was recorded over time for each member of the 48 families. This data set has excessive zeros as family members may not use health care facilities for a given period of time. Ma *et al.*[10] proposed a compound Poisson random effect approach to characterize both the excessive zeros and the rest of the data in an integral way. Their approach accounted for clustering effects at both family and subject levels, but ignored the serial dependence among observations over time. Hasan *et al.*[3] modeled the serial dependence among observations over time within a subject for subject-specific longitudinal count data with excessive zeros using a compound Poisson random effect. In this paper, we incorporate both clustering effects and serial dependence into the multilevel longitudinal zero-inflated count data. This approach merges population-averaged and cluster-specific analyses (see Ma and Jørgensen[11]). As in health care utilization data, count responses are collected longitudinally over six years. Like traditional models for longitudinal data, in this paper we assume serial dependence within a cluster with decreasing correlation over time.[3] For health care utilization data, Ma *et al.*[10] showed that after accounting for subject level covariates, the remaining clustering effects at the subject level beyond what can be explained at the family level are relatively low. That is why we incorporate serial dependence at the family level. Note that in longitudinal data, there may exist additional over-dispersion beyond that which can be captured by the serially dependent random effects (see Hasan et al.,[3] and Sutradhar and Bari[17]). Our proposed model can capture serial dependence, additional over-dispersion as well as the excessive zeros in the multilevel longitudinal count data.

After introducing the random effects Poisson models for multilevel longitudinal count data with excessive zeros in Section 2, we develop a quasi-likelihood approach for estimating the model parameters in Section 3. In Section 4, we use the proposed methodology to analyze health care utilization data presented in Ma et al.[10]

2. Zero-Inflated Poisson Mixed Model for Multilevel Longitudinal Data

In this section, we introduce an approach to model multilevel zero-inflated longitudinal count data that can unify population-averaged and cluster-specific analyses. To do this we incorporate Tweedie's compound Poisson distribution (see Jørgensen[4,5]) to model the effects on zeros and positive counts within a cluster. To accommodate serial correlation structure of the longitudinal responses, we also incorporate an AR(1) correlation structure. The ZIP mixed model for multilevel longitudinal data and the analogous moment structures are presented in subsequent subsections.

2.1. The Model

Let Y_{itj} represent the longitudinal count response recorded from the jth $(j = 1, 2, \ldots, n_i)$ subject at the tth $(t = 1, 2, \ldots, T)$ time point of the ith $(i = 1, 2, \ldots, K)$ independent family or cluster. Then the response vector can be expressed as $\boldsymbol{Y} = (Y_{111}, \ldots, Y_{11n_1}, \ldots, Y_{KT1}, \ldots, Y_{KTn_K})'$. We consider the family specific random effect u_{it} for the response at the tth time point of the ith family and the subject specific random effect v_{itj} for the response from the jth subject at the tth time point of the ith family. Let $\boldsymbol{W} = (\boldsymbol{U}, \boldsymbol{V})'$ denote the vector of the random effects where $\boldsymbol{U} = (\boldsymbol{U}_1, \ldots, \boldsymbol{U}_i, \ldots, \boldsymbol{U}_K)'$ with $\boldsymbol{U}_i = (U_{i1}, \ldots, U_{it}, \ldots, U_{iT})'$ and $\boldsymbol{V} = (\boldsymbol{V}_1, \ldots, \boldsymbol{V}_i, \ldots, \boldsymbol{V}_K)'$ with $\boldsymbol{V}_i = (\boldsymbol{v}_{i1}, \ldots, \boldsymbol{v}_{it}, \ldots, \boldsymbol{v}_{iT})'$ where $\boldsymbol{v}_{it} = (V_{it1}, \ldots, V_{itj}, \ldots, V_{itn_i})'$ respectively. Our model is based on the following three assumptions:

Assumption 1. First level random effects $U = (U_{11}, \ldots, U_{it}, \ldots, u_{KT})'$ are identically distributed, following an exponential distribution with rate λ. That is

$$E(U_{it}) = \frac{1}{\lambda}, \quad \text{var}(U_{it}) = \frac{1}{\lambda^2} \quad \text{and} \quad \text{corr}(U_{it}, U_{i't'}) = \begin{cases} \rho^{|t-t'|} & \text{if } i = i' \\ 0 & \text{if } i \neq i'. \end{cases}$$

Assumption 2. Given the first level random effects $U = u = (U_{11}, \ldots, U_{it}, \ldots, U_{KT})$, the second level random effects $V_{111}, \ldots, V_{itj}, \ldots, V_{KTn_K}$ are conditionally independent where the conditional distribution of V_{itj}, given $U = u$, depends on u_{it} only and follows the compound Poisson distribution, $\text{Tw}_q(u_{it}, \tau^2 u_{it}^{1-q})$ for $1 < q < 2$ in the notation of Ma and Jørgensen,[11] where q is the index parameter. More specifically, we have

$$V_{itj}|U \sim c_q(u_{it}; \tau^2) \exp\left\{ \frac{1}{\tau^2} \left(\frac{v_{itj}}{1-q} - \frac{u_{it}}{2-q} \right) \right\} \quad \text{where} \quad 1 < q < 2,$$

where the explicit expression for $c_q(y; \tau^2)$ is given by Jørgensen[5] but is immaterial in our derivation in the following sections except its positive probability mass at zero $Pr(V_{itj} = 0 \mid U) = \exp\left\{ \frac{-u_{it}}{\tau^2(2-q)} \right\}$, which can be denoted by $\phi(u_{it})$. The corresponding conditional mean and variance are given by $E(V_{itj}|U) = u_{it}$ and $\text{Var}(V_{itj}|U) = \tau^2 u_{it}$, which implies that the unconditional mean is $E(V_{itj}) = 1/\lambda$ and the unconditional covariance of v_{itj} and $v_{i't'j'}$ can be expressed as

$$\text{cov}(V_{itj}, V_{i't'j'}) = \begin{cases} \frac{\rho^{|t-t'|}}{\lambda^2} & \text{if } i = i'; t \neq t' \\ \frac{1+\lambda\tau^2}{\lambda^2} & \text{if } (i,t,j) = (i',t',j'). \\ 0 & \text{if } i \neq i' \end{cases}$$

Note that the unconditional moment structure of V is independent of the index parameter q.

Assumption 3. Given the random effects $W = w$, the components of Y are conditionally independent, and the conditional distribution of Y_{itj}, given $W = w$, depends on v_{itj} only, which is

$$Y_{itj} \mid W \sim \begin{cases} 0 & \text{with probability } \phi(u_{it}) \\ Y_{itj}^* \mid W & \text{with probability } 1 - \phi(u_{it}) \end{cases} \tag{1}$$

where $Y_{itj}^* \mid W \sim \text{Poisson}(\mu_{itj} v_{itj})$ and $\mu_{itj} = \exp(x_{itj}'\beta)$ with vector of covariates x_{itj} and regression parameter vector β. Thus the conditional expectation $E(Y_{itj} \mid W) = (1 - \phi(u_{it}))\mu_{itj} v_{itj}$ is smaller than $E(Y_{itj}^* \mid W)$ because of the mixture with zero. The mixture of zero and Poisson components here is modeled by a single subject random effects distribution, the compound Poisson distribution, in accordance with ZIP models. The mixture is done through the conditional probability $\phi(u_{it}) = Pr(V_{itj} = 0 \mid \boldsymbol{W})$ instead of the unconditional probability $Pr(V_{itj} = 0)$ to allow more flexibility in characterizing the heterogeneity among clusters (see Ma et al.[10]). Note that the random effects structure of this model is similar to that considered by Ma and Jørgensen,[11] Ma et al.[10] and Hasan et al.,[3] although incorporating both clustering effect and auto-correlation structure as well as the mixture with excessive zeros in (1) leads to a complex moment structure.

2.2. Mean and Covariance structure

The main focus of this section is to present the mean and covariance structures of the zero-inflated multilevel longitudinal Poisson mixed model discussed in the previous subsection. Following Hasan et al.,[3] the moments of the model can be obtained after some algebraic calculations by methods of conditioning on random effects. The unconditional mean and covariance are presented here to facilitate the parameter estimation.

From Assumption 3, the unconditional expectation of the response vector $\boldsymbol{Y}$ can be expressed as

$$E(\boldsymbol{Y}) = \boldsymbol{\mu} \left[\frac{1}{\lambda} - \frac{\lambda}{\left(\lambda + \frac{1}{\tau^2(2-q)}\right)^2} \right] = \boldsymbol{\mu}\eta_1(\lambda, \tau^2), \qquad (2)$$

where

$$\eta_1(\lambda, \tau^2) = \left[\frac{1}{\lambda} - \frac{\lambda}{\left(\lambda + \frac{1}{\tau^2(2-q)}\right)^2} \right]. \qquad (3)$$

Here $E(Y_{itj}) = E\{E(Y_{itj}|\boldsymbol{W})\} = E[\{1 - \phi(U_{it})\}\mu_{itj}V_{itj}] = \mu_{itj}P(Y_{itj} = 0)$; therefore our approach merges population-averaged and cluster-specific analyses.[8,11] The unconditional covariance i.e. $\text{Cov}(\boldsymbol{Y})$ is a $\left(T \times \sum_{i=1}^{K} n_i\right) \times \left(T \times \sum_{i=1}^{K} n_i\right)$ block diagonal matrix where the ith block, $\boldsymbol{B}_i$, can be expressed as

$$\boldsymbol{B}_i = \boldsymbol{\mu}_i \boldsymbol{D}_i \boldsymbol{\mu}_i' + \text{diag}\left[\boldsymbol{\mu}_i^2 \left\{\eta_2(\lambda, \tau^2) + \tau^2\eta_1(\lambda, \tau^2)\right\}\right] \qquad (4)$$

$$+ \text{diag}\left\{\boldsymbol{\mu}_i\eta_1(\lambda, \tau^2)\right\}, \qquad (5)$$

where $\boldsymbol{\mu}_i = (\mu_{i11}, \ldots, \mu_{i1n_i}, \ldots, \mu_{iT1}, \ldots, \mu_{iTn_i})'$ and $\boldsymbol{\mu}_i^2 = (\mu_{i11}^2, \ldots, \mu_{i1n_i}^2, \ldots, \mu_{iT1}^2, \ldots, \mu_{iTn_i}^2)'$. In (5), $\eta_1(\lambda, \tau^2)$ is as defined in (3), $\eta_2(\lambda, \tau^2)$ can be expressed as

$$\eta_2(\lambda, \tau^2) = \left[\frac{2\lambda}{\left(\lambda + \frac{1}{\tau^2(2-q)}\right)^3} - \frac{2\lambda}{\left(\lambda + \frac{2}{\tau^2(2-q)}\right)^3} \right] \tag{6}$$

and D_i is a $(Tn_i) \times (Tn_i)$ matrix defined as

$$D_i = \begin{bmatrix} D(\rho^0) & D(\rho^1) & D(\rho^2) & \ldots \ldots D(\rho^{T-2}) & D(\rho^{T-1}) \\ D(\rho^1) & D(\rho^0) & D(\rho^1) & \ldots \ldots D(\rho^{T-3}) & D(\rho^{T-2}) \\ \vdots & \vdots & \vdots & \vdots & \vdots & \vdots \\ \vdots & \vdots & \vdots & \vdots & \vdots & \vdots \\ D(\rho^{T-2}) & D(\rho^{T-3}) & D(\rho^{T-4}) & \ldots \ldots D(\rho^0) & D(\rho^1) \\ D(\rho^{T-1}) & D(\rho^{T-2}) & D(\rho^{T-3}) & \ldots \ldots D(\rho^1) & D(\rho^0) \end{bmatrix}, \tag{7}$$

where $D(\rho^{|t-t'|}) = \eta_3(\lambda, \tau^2, \rho^{|t-t'|})\mathbf{1}_{n_i}\mathbf{1}'_{n_i}$ with $\eta_3(\lambda, \tau^2, \rho^{|t-t'|})$ as

$$\eta_3(\lambda, \tau^2, \rho^{|t-t'|}) = \frac{\rho^{|t-t'|}}{\lambda^2} + \frac{1 + \rho^{|t-t'|}}{\left(\lambda + \frac{1}{\tau^2(2-q)}\right)^2} - \frac{2\lambda\rho^{|t-t'|}}{\left(\lambda + \frac{1}{\tau^2(2-q)}\right)^3} - \frac{\lambda^2}{\left(\lambda + \frac{1}{\tau^2(2-q)}\right)^4}$$

$$-2\lambda\rho^{|t-t'|} \left\{ \frac{1}{\left(\lambda + \frac{\rho^{|t-t'|}}{\tau^2(2-q)}\right)^3} - \frac{1}{\left(\lambda + \frac{1+\rho^{|t-t'|}}{\tau^2(2-q)}\right)^3} \right\}$$

$$\times \left\{ \rho^{|t-t'|} + \frac{\lambda(1 - \rho^{|t-t'|})}{\left(\lambda + \frac{1}{\tau^2(2-q)}\right)} \right\} - \frac{\lambda^2(1 - \rho^{|t-t'|})}{\left(\lambda + \frac{1}{\tau^2(2-q)}\right)^2}$$

$$\times \left\{ \frac{1}{\left(\lambda + \frac{\rho^{|t-t'|}}{\tau^2(2-q)}\right)^2} - \frac{1}{\left(\lambda + \frac{1+\rho^{|t-t'|}}{\tau^2(2-q)}\right)^2} \right\}. \tag{8}$$

In the next section we use the above expressions to estimate the regression and random effects parameters.

3. Parameter Estimation

In this section, we discuss the estimation of the regression parameter β and the random effect parameters λ, τ^2 and ρ. We use a quasi-likelihood approach for estimation of the regression parameter whereas the method of moments is used for estimating the random effect parameters.

3.1. *Estimation of regression parameters*

We first present the quasi-likelihood (QL) estimation approach for estimating the regression parameters under the assumption that the random effects parameters are known. We will relax this assumption and discuss the estimation technique for the random effect parameters in next subsection. The QL estimating equation for the regression parameter β can be denoted by $\psi(\beta)$, that is expressed as

$$\psi(\beta) = \sum_{i=1}^{K} X_i'\text{diag}\{E(\mathbf{Y}_i)\}\,\text{cov}^{-1}(\mathbf{Y}_i)[\mathbf{Y}_i - E(\mathbf{Y}_i)] = 0 \qquad (9)$$

where the ith component corresponds to the ith family or cluster where $\text{diag}\{E(\mathbf{Y}_i)\}$ are the corresponding diagonal matrices of vectors $E(\mathbf{Y}_i)$, respectively. Without loss of generality, we assume that the matrix of the covariates $X_i' = (x_{i11}, \ldots, x_{iTn_i})$ is of full rank. Following Ma and Jørgensen,[11] under mild regularity consistent and asymptotically normal with asymptotic mean β and asymptotic variance given by the inverse of the sensitivity matrix $S(\beta) = E_\beta\{\partial\psi(\beta)/\partial\beta\}$ as the subjects are assumed to be independent. In addition, the estimating estimating $\psi(\beta) = 0$ is optimal in the sense that it attains the minimum asymptotic covariance for the estimator of β within the class of all linear functions of Y.[11] As in Ma et al.,[10] this estimation equation $\psi(\beta) = 0$ can be solved iteratively using the scoring algorithm, where the value of β is updated following

$$\beta^* = \beta - S^{-1}(\beta)\psi(\beta),$$

with the explicit expression of the sensitivity matrix is given by

$$S(\beta) = X'\text{diag}\{E(\mathbf{Y})\}\,\text{cov}^{-1}(\mathbf{Y})\text{diag}\{E(\mathbf{Y})\}\,X$$

with $X' = (x_{111}, \ldots, x_{KTn_T})$ and $E(\mathbf{Y}) = \{E(Y_{111}), \ldots, E(Y_{KTn_T})\}'$.

3.2. *Estimation of random effects parameters*

To estimate the regression parameter β in the previous subsection we assumed the random effect parameters are known. In this section, we present a moment approach to estimate the unknown random effect parameters λ, τ^2 and ρ based on three unbiased estimating equations, which are obtained using the mean, variance and covariance structures, respectively.[3,10] The first one employs the mean structure:

$$\xi_1(\lambda, \tau^2) = \frac{1}{n}\sum_{i=1}^{K}\sum_{t=1}^{T}\sum_{j=1}^{n_i}[y_{itj} - E(Y_{itj})] = \frac{1}{n}\sum_{i=1}^{K}\sum_{t=1}^{T}\sum_{j=1}^{n_t}\left[y_{itj} - \mu_{itj}\eta_1(\lambda, \tau^2)\right],$$

where $n = T * \sum_{i=1}^{K} n_i$. The second one is based on the variance structure which can be expressed as

$$\xi_2(\lambda, \tau^2) = \frac{1}{n} \sum_{i=1}^{K} \sum_{t=1}^{T} \sum_{j=1}^{n_i} \left[\{y_{itj} - E(Y_{itj})\}^2 - Var(Y_{itj}) \right]$$

$$= \frac{1}{n} \sum_{i=1}^{K} \sum_{t=1}^{T} \sum_{j=1}^{n_i} \left[\left\{ y_{itj} - \mu_{itj} \eta_1(\lambda, \tau^2) \right\}^2 - \left\{ \mu_{itj}^2 \left(\eta_3(\lambda, \tau^2, 1) \right. \right. \right.$$

$$\left. \left. \left. + \eta_2(\lambda, \tau^2) + \tau^2 \eta_1(\lambda, \tau^2) \right) + \mu_{itj} \eta_1(\lambda, \tau^2) \right\} \right].$$

The third equation is based on the covariance structure, which can be expressed as

$$\xi_3(\lambda, \tau^2, \rho) = \frac{1}{\sum_{i=1}^{K} \sum_{t=1}^{T-1} n_i} \sum_{i=1}^{K} \sum_{t=1}^{T-1} \sum_{j=1}^{n_i} \left[\{ (y_{itj} - E(Y_{itj})) \right.$$

$$\left. \times (y_{i(t+1)j} - E(Y_{i(t+1)j})) \} - \text{cov}(Y_{itj}, Y_{i(t+1)j}) \right]$$

$$= \frac{1}{\sum_{i=1}^{K} \sum_{t=1}^{T-1} n_i} \sum_{i=1}^{K} \sum_{t=1}^{T-1} \sum_{j=1}^{n_i} \left[\left\{ y_{itj} - \mu_{itj} \eta_1(\lambda, \tau^2) \right\} \right.$$

$$\left. \times \left\{ y_{i(t+1)j} - \mu_{i(t+1)j} \eta_1(\lambda, \tau^2) \right\} - \mu_{itj} \mu_{i(t+1)j} \eta_3(\lambda, \tau^2, \rho) \right].$$

The explicit forms for the estimators of λ, τ^2 and ρ are not straightforward. However, with given regression parameter estimators, the estimators of λ, τ^2 and ρ can be obtained by solving the estimating equations $\xi_1(\lambda, \tau^2) = 0$, $\xi_2(\lambda, \tau^2) = 0$ and $\xi_3(\lambda, \tau^2, \rho) = 0$ simultaneously. To do that we use optimization techniques in the free statistical software R with the function optimize to solve these three estimating equations numerically. Taking the regression parameter estimates obtained from standard Poisson models as initial values for regression parameter estimates, the algorithm then iterates between updating random effects parameter estimates via the R function optimize and updating the regression parameter estimates via the scoring algorithm.

4. Application

We have applied our proposed methodology to the analysis of health care utilization data which were collected from St. John's, Newfoundland, Canada and consist of 48 families. These families were observed for six consecutive years between 1985 and 1990. Each year the number of annual visits to health care was recorded for each member of every family, along with covariate information. Out of these 48 families, 36 of the families had 4 members and 12 of the families had 3 members. In other words, we have 4 count responses from 36 families and 3 responses from 12 families for six years. Around twenty five percent of these recorded observations were zeros, which is quite high. Covariate information on sex, education level, number of chronic conditions and age (in years) was also collected at the beginning of the study. In this study our scientific interest is to investigate if there are any significant differences in health care utilization due to any of the covariates after accounting for family and individual effects as well as longitudinal correlation. We considered a family level exponential AR(1) auto-correlation structure, since Hasan *et al.*[3] show that longitudinal responses with excessive zeros usually incorporate an exponentially distributed shape.

We consider the annual numbers of visits to health care repeated for six years by each individual of every family as the response variable in our model defined in Section 2. As covariates we took Sex (1 for female, 0 for male), Education (1 for less than high school, 0 otherwise), the logarithm of the number of chronic conditions (Chronic) and logarithm of age (Age). To avoid problems with zeros for the covariate Chronic we added 0.1 before taking the logarithm. Thus the Poisson mean parameter can be expressed as

$$\mu_{itj} = \exp\{\beta_0 + \beta_1 \text{Sex} + \beta_2 \text{Education} + \beta_3 \text{Chronic} + \beta_4 \text{Age}\}, \quad (10)$$

where $i = 1, \ldots, 48$, $t = 1, \ldots, 6$ and $j = 1, \ldots, n_i$ where $n_i = 3$ or 4 representing the members in the ith family. We modeled the subject random effects using Tweedie's compound Poisson distribution with index parameter $p = 1.5$. The estimates and the corresponding standard errors of the regression parameters are presented in Table 1. We investigated if interaction terms between Sex and each of Age and Chronic were needed. However, we found these interaction terms were insignificant, so they are omitted. The program took 50 seconds to complete the data analysis with four decimal point precision. The estimates of the random effects parameters are also displayed in Table 1.

We see in Table 1 that Sex, Chronic and Age have significant effects and Education has an insignificant effect at the 5% significance level. The results show that the health care utilization increases as age and number of chronic conditions increase. The monotonic trend with the age variable here is not a surprise since all the subjects were adults aged between 19.2 and 85.2 at the beginning of the study in 1985. Although the oldest subject

Table 1. Estimates of parameters, standard errors for health care utilization data.

Coefficient	$q = 1.1$ Est.[a]	Sd. Er.[b]	$q = 1.5$ Est.	Sd. Er.	$q = 1.9$ Est.	Sd. Er.
Intercept	0.8347	0.4090	0.8252	0.3844	0.8513	0.3697
Sex	0.5135	0.0801	0.5147	0.0755	0.5157	0.0708
Education	-0.1049	0.1333	-0.1061	0.1251	-0.1076	0.1169
Chronic	0.1404	0.0344	0.1405	0.0324	0.1407	0.0303
Age	0.3818	0.1223	0.3822	0.1150	0.3807	0.1077
λ	3.4173		3.5033		3.6173	
τ^2	0.0981		0.0969		0.0901	
ρ	0.2408		0.2622		0.2812	

Note: [a] Estimate. [b] Standard Error.

reached 91 years old at the end of the study, there were no missing values in the data set due to death. Women are also likely to have higher health care utilization than men.

The estimates of λ, τ^2 and ρ are 3.5033, 0.0969 and 0.2622, respectively, using the index parameter $q = 1.5$. This indicates the temporal random effect has mean $1/\lambda = 0.2854$ and variance $1/\lambda^2 = 0.0815$ with auto-correlation of $\rho = 0.2622$. The longitudinal correlation of 0.2622 indicates that the number of heath care visits in a particular year, to some extent, depends on that of the previous year. An estimate of $\tau^2 = 0.0969$ indicates that there is more variation beyond what can be characterized by family level temporal random effect. To assess the influence of the index parameter q in the choice of the compound Poisson distribution on our analysis results, as in Ma et al.,[10] we fitted the model specified in (10) with $q = 1.1, 1.2, 1.3, 1.4, 1.6, 1.7, 1.8$ and 1.9 as well. The estimates of both the regression and random effects parameters remained similar across these choices of q. The results are thus illustrated with the two extreme cases of $q = 1.1$ and 1.9 together with $q = 1.5$ in Table 1.

5. Discussion

In this paper, we have introduced a Zero-inflated Poisson mixed model to analyze multilevel longitudinal count data with excessive zeros. We presented a unified approach between population-averaged and cluster-specific analyses for modelling zero-inflated multilevel longitudinal count responses, as the marginal mean of the response vector can be obtained by taking the expectations on the random effects of the conditional mean of the responses; both the marginal and conditional means have simple linear structures.

This is also true for the models introduced in Ma *et al.*[10] and Hasan *et al.*[3] The equal number of time points per cluster was assumed in our model for notational convenience. However, neither the moment structure nor the quasi-likelihood approach depends on this assumption. Another advantage of our quasi-likelihood approach is its computational efficiency with analysis of very large data sets (see Ma *et al.*[12]). Following Ma *et al.*[10] and Hasan *et al.*,[3] our procedure for estimating model parameters, including components of complex covariance structures, is efficient and viable to implement using optimization methods.

We chose two particular random effects distributions, exponential AR(1) for the family effect and longitudinal correlation structure at the family level and compound Poisson for subject effects with index parameter $q = 1.5$ to analyze the health care utilization data. As in Ma *et al.*[10] and Hasan *et al.*,[3] our inference on the fixed effects is robust to the index parameter q for compound Poisson random effects. This is probably because the first two moments of our random effects do not depend on the shape parameter q. As noted by Neuhaus and McCulloch,[16] with the first two moments of the random effects given, the shape of the random effects distribution does not have much affect on the inference of the fixed effects.

Acknowledgments

This research was partially supported by grants from the Natural Sciences and Engineering Research Council of Canada. The authors thank two referees whose comments led to substantial improvements in the paper.

References

1. D. B. Hall, *Biometrics* **56**, 1030 (2000).
2. M. T. Hasan and G. Sneddon, *Communications in Statistics: Simulation and Computation* **38**, 638 (2009).
3. M. T. Hasan, G. Sneddon and R. Ma, *Biometrical Journal* **51**, 946 (2009).
4. B. Jørgensen, *Journal of Royal Statistical Society, Series B* **49**, 127 (1987).
5. B. Jørgensen, *The Theory of Dispersion Models* (Chapman and Hall, London, 1997).
6. K. F. Lam, H. Xue and Y. B. Cheung, *Biometrics* **62**, 996 (2002).
7. D. Lambert, *Technometrics* **34**, 1 (1992).
8. Y. Lee and J. A. Nelder, *Journal of Royal Statistical Society, Series B* **58**, 619 (1996).
9. S. Lu, Y. Lin and W. J. Shih, *Biometrics* **60**, 257 (2004).
10. R. Ma, M. T. Hasan and G. Sneddon, *Statistics in Medicine* **28**, 2356 (2009).
11. R. Ma and B. Jørgensen, *Journal of Royal Statistical Society, Series B* **69**, 625 (2007).

12. R. Ma, D. Krewski and R. T. Burnett, *Biometrika* **90**, 157 (2003).
13. Y. Min and A. Agresti, *Journal of Iranian Statistical Society* **1**, 7 (2002).
14. Y. Min and A. Agresti, *Statistical Modelling* **5**, 1 (2005).
15. J. Mullahy, *Journal of Econometrics* **33**, 341 (1986).
16. J. M. Neuhaus and C. E. McCulloch, *Journal of Royal Statistical Society, Series B* **68**, 859 (2006).
17. B. C. Sutradhar and W. Bari, *Sankhyā, The Indian Journal of Statistics* **69**, 671 (2007).
18. J. M. Ver Hoef and J. K. Jansen, *Environmetrics* **18**, 697 (2007).
19. K. K. W. Yau and A. H. Lee, *Statistics in Medicine* **20**, 2907 (2001).

META ANALYSIS OF BINARY OUTCOMES DATA IN CLINICAL TRIALS

PERLA SUBBAIAH

Department of Mathematics & Statistics, Oakland University,
Rochester, MI 48309, USA
E-mail: perla@oakland.edu

AVISHEK MALLICK

Department of Mathematics & Statistics, University of New Hampshire
Durham, NH 03824, USA
E-mail: am251081@gmail.com

TUSAR K. DESAI

William Beaumont Hospital
Royal Oak, MI 48073, England
E-mail: tusardesai@aol.com

The primary objective of this paper is to present different methods for combining the results of several investigations involving binary outcomes. Typically the studies involve comparison of a treatment with control. However, in this study we focus on only non-comparative binary outcomes. Towards this end, we include methods based on different effect sizes, which are functions of the success rate of the attribute of interest, namely, (i) success rate $\hat{p}$, (ii) log of success rate $\ln(\hat{p})$, (iii) log of odds ratio $\ln(\frac{\hat{p}}{1-\hat{p}})$, and (iv) variance stabilizing transformation of the success rate $\sin^{-1}(\sqrt{\hat{p}})$. These meta analytic methods are illustrated using two different data sets: a data set based on pharmacoeconomic applications using antifungal onychomycosis lacquers, where the success rate of ciclopirox lacquer is relatively large and another data set involving cancer incidence, which is a rare event. The meta analytic methods behave differently depending on which type of data is used. When the data set involves rare events, application of the variance stabilizing transformation can be carried out without adjusting the success rate and its standard error, where as the other methods do require that adjustment, and hence the variance stabilizing transformation is preferred for the meta analysis.

1. Introduction

Meta analysis is extensively used by health researchers in order to combine the information from several investigations to assess the effectiveness of a treatment and update their knowledge. Sutton et al.[18] while describing various advantages of meta analysis, mention that "... it is the increased power gained from synthesizing results from different studies that makes the systematic review an important tool". One of the earlier books on statistical methods for meta analysis is Hedges and Olkin(1985).[14] There are several books and articles related to the recent developments in meta analysis namely Sutton et al.,[18] Sutton and Higgins,[19] Hartung, Knapp and Sinha[13] and Borenstein et al.[2] In this paper we consider methods of combining non-comparative studies with binary data. However, these methods can be used for comparative studies with treatment and control groups.

Suppose p is the proportion of the attribute of interest in the population. Some commonly used measures of the effect size are $\theta_1 = p$, $\theta_2 = \ln(\frac{p}{1-p}) = logit(p)$, $\theta_3 = \ln(p)$, and Fisher's variance stabilizing transformation $\theta_4 = \sin^{-1}(\sqrt{p})$. Suppose a random sample of size n is taken from the population, and X is the number of subjects with the attribute of interest. Then $\hat{p} = X/n$ is an estimate of p, and the corresponding estimates of the effect size θ are $\hat{\theta}_1 = \hat{p}$, $\hat{\theta}_2 = \ln(\frac{\hat{p}}{1-\hat{p}})$, $\hat{\theta}_3 = \ln(\hat{p})$ and $\hat{\theta}_4 = \sin^{-1}(\sqrt{\hat{p}})$. The variances of these estimates are given by $Var(\hat{\theta}_1) = \frac{p(1-p)}{n}$, $Var(\hat{\theta}_2) = \frac{(1-p)}{np}$, $Var(\hat{\theta}_3) = \frac{1}{np(1-p)}$, and $Var(\hat{\theta}_4) = \frac{1}{4n}$.

In this paper we discuss methods of estimating p based on these measures of effect size. In Section 2, we discuss the confidence intervals for p based on the assumption of Binomial distribution. In Section 3, we describe the meta analysis methods using these four effect sizes and estimate of p with fixed effects models and random effects models. These methods are illustrated using two sets of data. In Section 4, we compare the methods using data based on antifungal onychomycosis lacquers, where the success rate of ciclopirox lacquer is relatively large. In Section 5, a data set involving the incidence of esophageal cancer in Barrett's esophagus, which is a rare event, is considered. In Section 6, we discuss appropriate modifications of the methods when each study involves both treatment and control groups.

2. Confidence Intervals for the Proportion of Interest

In this section we describe some commonly used methods for obtaining the confidence interval for the population proportion p. Suppose X is the number of "successes" in n independent trials. This could be interpreted as follows: For example, n may indicate the number of subjects and X may be the number of subjects with the characteristic of interest. In another setting, n may be the total number of person-years observed and X may be the number of incident cancers. Let $X=\sum_{i=1}^{n} X_i$, where X_i, $i = 1, 2, ..., n$ are independent identically distributed as Bernoulli random variables with parameter p. Also, let $\hat{p}=\frac{X}{n}$ denote the unbiased estimator of p.

Assuming that X follows Binomial distribution with parameters n and p, the following intervals may be considered:

(a) An exact confidence interval for p with confidence level $100(1 - \alpha)\%$ is (see Casella and Berger[3])

$$\frac{1}{1 + \frac{n-x+1}{x} F_{2(n-x+1),2x;\frac{\alpha}{2}}} \leq p \leq \frac{\frac{x+1}{n-x} F_{2(x+1),2(n-x);\frac{\alpha}{2}}}{1 + \frac{x+1}{n-x} F_{2(x+1),2(n-x);\frac{\alpha}{2}}}, \quad (1)$$

where $F_{\nu_1,\nu_2;\alpha}$ is the upper α percentile of an F distribution with (ν_1,ν_2) degrees of freedom. The end point adjustment is made so that if $x = 0$, the lower end point is 0, and if $x = n$, the upper end point is 1. These intervals are due to Clopper and Pearson.[5]

(b) The confidence interval based on the normal approximation of the binomial distribution (also known as Wald interval), is

$$\hat{p} \pm z_{\alpha/2} \sqrt{\frac{\hat{p}(1 - \hat{p})}{n}}. \quad (2)$$

(c) The interval for p based on the score method, i.e., by inverting the score statistic $(\hat{p} - p)/\sqrt{p(1 - p)/n}$ is

$$\frac{(2\hat{p} + z_{\alpha/2}^2/n) \pm \sqrt{(2\hat{p} + z_{\alpha/2}^2/n)^2 - 4\hat{p}^2(1 + z_{\alpha/2}^2/n)}}{2(1 + z_{\alpha/2}^2/n)}. \quad (3)$$

(d) The interval based on Fisher's variance stabilizing transformation, as suggested by Chen[4] is

$$\sin^{-1}(\sqrt{p}) \in \sin^{-1}(\sqrt{\hat{p}}) \pm z_{\alpha/2} \frac{1}{\sqrt{4n}} = (L, U)$$

$$i.e, p \in (\sin^2(L), \sin^2(U)). \quad (4)$$

We illustrate these intervals with a few selected values for X and n as given in Table 1. While comparing the confidence intervals, both the expected length of the intervals and the coverage probabilities need to be considered. In Table 1, the results of a simulation study are given. A sample

Table 1. Coverage Probabilities & Expected Lengths of 95% Confidence Intervals for p.

Method	$n = 20$ p			$n = 100$ p			$n = 500$ p		
	0.05	0.3	0.65	0.05	0.3	0.65	0.05	0.3	0.65
Exact Interval	.974	.976	.969	.983	.963	.956	.961	.954	.956
	(.235)	(.413)	(.428)	(.094)	(.186)	(.194)	(.040)	(.082)	(.085)
Wald's Interval	.998	.950	.938	.875	.950	.940	.931	.949	.952
	(.156)	(.388)	(.406)	(.082)	(.179)	(.186)	(.038)	(.080)	(.084)
Score Interval	.923	.976	.969	.966	.940	.956	.952	.954	.946
	(.219)	(.365)	(.377)	(.088)	(.176)	(.183)	(.038)	(.080)	(.083)
VS Interval	.628	.950	.938	.953	.950	.956	.947	.949	.952
	(.159)	(.377)	(.393)	(.082)	(.177)	(.185)	(.038)	(.080)	(.083)

of 10,000 values are generated from a binomial distribution with parameters n and p. The values considered for the parameters are: $n = 20, 100, 500$ and $p = 0.05, 0.30, 0.65$. For each value of X obtained, the intervals in equations 1-4 are calculated, and the length of the interval as well as whether each interval contained the parameter p is noted. Then, the average of the lengths of the intervals, and the proportion of the intervals containing the parameter p are obtained. We conclude the following from the results given in Table 1:

(i) When n is large, it seems that the expected length of the confidence interval (4), which is based on variance stabilizing transformation is shortest or very close to shortest among the four methods considered, and the coverage probability is also close to the nominal level.

(ii) When p is small, the same observation holds with the exception of $n = 20$.

3. Meta Analysis of Binary Outcome Data

In this section we describe meta analysis of several studies with binary outcome data. We are interested in estimating p, the proportion of the attribute of interest in the population. We study various measures of the effect size $\theta_1 = p$, $\theta_2 = \ln(p/1 - p) = logit(p)$, $\theta_3 = \ln(p)$ and Fisher's variance stabilizing transformation, $\theta_4 = \sin^{-1}(\sqrt{p})$. Suppose a random sample of size n is taken from the population, and X is the number of subjects with the attribute of interest. Then $\hat{p} = X/n$ is an estimate of p, and the corresponding estimates of the effect size θ are $\hat{\theta}_1 = \hat{p}$, $\hat{\theta}_2 = \ln(\hat{p}/1-\hat{p})$, $\hat{\theta}_3 = \ln(\hat{p})$ and $\hat{\theta}_4 = \sin^{-1}(\sqrt{\hat{p}})$. The variances of these estimates are given by $Var(\hat{\theta}_1) = \frac{p(1-p)}{n}$, $Var(\hat{\theta}_2) = \frac{(1-p)}{np}$, $Var(\hat{\theta}_3) = \frac{1}{np(1-p)}$, and $Var(\hat{\theta}_4) = \frac{1}{4n}$.

In order to describe the meta analysis method, we use the following notation. Let k denote the number of studies, and T_i is the estimate of effect size θ_i from i^{th} study, $i = 1, 2, ..., k$ and $\sigma^2(T_i)$ and $\hat{\sigma}^2(T_i)$ are the variance and estimated variance of T_i respectively. The meta analysis involves testing the null hypothesis of equal effect size in all k studies, i.e., $\theta_1 = \theta_2 = ... = \theta_k = \theta$, where θ is the common effect size.

We can estimate the common effect size θ by taking a weighted average of T_i's. The estimate, due to Cochran,[6] is

$$\hat{\theta} = \frac{\sum_{i=1}^{k} w_i T_i}{\sum_{i=1}^{k} w_i}, \tag{5}$$

where $w_i = 1/\sigma^2(T_i)$, which minimizes $var(\hat{\theta})$ assuming T_i is an unbiased estimate of θ, and the weights w_i are known. In practice, if the weights involve unknown parameters, they are replaced by the estimates of the parameters , i.e., $w_i = 1/\hat{\sigma}^2(T_i)$ and are treated as known fixed quantities. It can be shown that

$$\hat{\sigma}^2 = \frac{1}{\sum_{i=1}^{k} w_i}. \tag{6}$$

A confidence interval for θ with confidence level $100(1-\alpha)\%$ can be obtained as

$$\hat{\theta} \pm z_{\alpha/2} \sqrt{\frac{1}{\sum_{i=1}^{k} w_i}}. \tag{7}$$

Once θ is estimated, an inverse transformation can then be used to obtain a combined estimate of p, depending on the effect size used.

The homogeneity of the effect sizes $H_0 : \theta_1 = \theta_2 = ... = \theta_k = \theta$ can be tested using the statistic

$$Q = \sum_{i=1}^{k} w_i(T_i - \hat{\theta})^2 = \sum_{i=1}^{k} w_i T_i^2 - \frac{(\sum_{i=1}^{k} w_i T_i)^2}{\sum_{i=1}^{k} w_i}. \tag{8}$$

Under $H_0 : \theta_1 = \theta_2 = ... = \theta_k = \theta$, Q is approximately distributed as χ^2_{k-1}. The null hypothesis of homogeneity of effect sizes is rejected if $P - value = P(\chi^2_{k-1} > q)$ is small, where q is the calculated value of Q, and χ^2_{k-1} is the chi-square random variable with $(k - 1)$ d.f.

Random Effects Model and Estimation of Between Study Variance τ^2 : In the notation of ANOVA model, consider $T_i|\theta_i \sim N(\theta_i, \sigma_i^2)$, and $\theta_i \sim N(\theta, \tau^2)$ for $i = 1, 2, ..., k$, and $T_1, T_2, ..., T_k$ are independent. This implies that the marginal distribution of T_i is $N(\theta, \sigma_i^2 + \tau^2)$. The most commonly used method of estimating τ^2 is due to DerSimonian and Laird[7] which uses Q statistic in (8). It can be shown that with the assumptions of the random effects model,

$$E(Q) = (n-1) + \left(\sum w_i - \frac{\sum w_i^2}{\sum w_i} \right) \tau^2. \tag{9}$$

Replacing $E(Q)$ with Q and solving for τ^2 leads to the DerSimonian and Laird estimate

$$\hat{\tau}^2 = max \left(0, \frac{Q - (n-1)}{\left(\sum w_i - \frac{\sum w_i^2}{\sum w_i} \right)} \right). \tag{10}$$

The estimate of θ that incorporates this estimate of τ^2 is

$$\hat{\theta}^* = \frac{\sum_{i=1}^{k} w_i^* t_i}{\sum_{i=1}^{k} w_i^*}, \tag{11}$$

where $w_i^* = 1/(\sigma_i^2 + \hat{\tau}^2)$. Similarly, the confidence interval for θ with confidence level $100(1-\alpha)\%$ based on the random effects model is

$$\hat{\theta}^* \pm z_{\alpha/2} \sqrt{\frac{1}{\sum_{i=1}^{k} w_i^*}}. \tag{12}$$

Remark 1: Some authors have suggested use of the random effects model to estimate the summary effect θ. If τ^2 is not significant, then the random effects model gives results very similar to that of the fixed effects model. Moreover, the test for the significance of τ^2 based on the statistic Q in (8) has low power, and even if the test based on Q is insignificant, it may not imply that the fixed effects model is appropriate.

Remark 2: Follmann and Proschan[11] have suggested an ad hoc method of obtaining the confidence interval for θ replacing the percentile from normal distribution with that of t-distribution, i.e.,

$$\hat{\theta}^* \pm t_{k-1;\alpha/2} \sqrt{\frac{1}{\sum_{i=1}^{k} w_i^*}}. \tag{13}$$

Hartung and Knapp[12] provided an improved confidence interval for θ,

$$\hat{\theta}^* \pm t_{k-1;\alpha/2} \sqrt{q} \tag{14}$$

where $q = \frac{\sum_{i=1}^{k} w_i^* (t_i - \hat{\theta}^*)^2}{(k-1) \sum_{i=1}^{k} w_i^*}$. They have investigated the performance of these methods through simulation studies, and found that the control of significance level or confidence level is better with the refined methods.

Remark 3: Jackson, Bowden and Baker[16] mention that "unless all studies are of similar size, this (DerSimonian and Laird procedure) is inefficient when estimating the between-study variance, but is remarkably efficient when estimating the treatment effect. If formal inference is restricted to statements about the treatment effect, and the sample size is large, there is little point in implementing more sophisticated methodology."

Estimation of τ^2, and I^2: The distribution of $\hat{\tau}^2$ is not even symmetric and $\hat{\tau}^2 \pm z_{\alpha/2} s.e(\hat{\tau}^2)$ will not provide accurate confidence interval for τ^2 unless k is very large. A simple method for confidence interval for τ^2 is as follows (Borenstein et $al.^2$).
Let

$$
B = \begin{cases}
\left[\frac{0.5\{\ln(Q) - \ln(k-1)\}}{\sqrt{2Q} - \sqrt{2(k-1)-1}} \right] & , if Q > k \\
\left[2(k-2)\left(1 - \frac{1}{3(k-2)^2}\right) \right]^{-0.5} & , if Q \le k
\end{cases}
$$

and $C = \left(\sum w_i - \frac{\sum w_i^2}{\sum w_i} \right)$.
Also, let

$$
L = \exp\{0.5\ln(Q/(k-1)) - z_{\alpha/2}B\},
$$

$$
U = \exp\{0.5\ln(Q/(k-1)) + z_{\alpha/2}B\}. \tag{15}
$$

Then the $100(1-\alpha)\%$ confidence interval for τ^2 is given by

$$
(LL_{\tau^2}, UL_{\tau^2}), \tag{16}
$$

where $LL_{\tau^2} = (k-1)(L^2 - 1)/C$, and $UL_{\tau^2} = (k-1)(U^2 - 1)/C$. Any value $LL_{\tau^2}, UL_{\tau^2}, \tau^2$ that is computed as less than zero, is set to be zero.

The proportion of the observed variance reflecting real differences in effect sizes is defined as

$$
I^2 = \left(\frac{Q - (k-1)}{Q} \right) 100\%.
$$

I^2, a percentage of total variation across studies that is due to heterogeneity rather than chance, is proposed by Higgins et $al.^{15}$ $I^2 = 0\%$ indicates no observed heterogeneity, and larger values show increasing heterogeneity. I^2 can also be interpreted as a ratio of "between variation" and "total variation". This also depends on the choice of effect measure that is used. I^2 is preferable to a test for heterogeneity in judging consistency of evidence, as suggested by Higgins et $al.^{15}$

The confidence interval for I^2 is given by

$$(LL_{I^2}, UL_{I^2}), \tag{17}$$

where $LL_{I^2} = \frac{L^2-1}{L^2}100\%$ and $UL_{I^2} = \frac{U^2-1}{U^2}100\%$, L and U are as given in (15). Any value LL_{I^2}, UL_{I^2}, I^2 that is computed as less than zero, is set to be zero.

Remark 4: Knapp, Biggerstaff and Hartung[17] proposed some tests of hypotheses and confidence interval methods for the heterogeneity parameter τ^2. They have illustrated the methods with two real data examples and evaluated through simulation studies.

Remark 5: Publication bias becomes an issue because only the studies with significant results are likely to be published. The results from smaller studies vary widely around the mean effect due to large random error. In comparative studies, a plot of sample size vs. treatment effect from the studies should be shaped like a funnel if there is no publication bias (Sutton et al.[18]). Funnel plot is an informal method to judge the potential bias, and is subjective in its interpretation. Begg and Majumdar[1] proposed a rank correlation test between effect estimates and their variances. Egger et al.[9] proposed a test based on the regression of the standardized effect estimates on the reciprocal of sample standard deviation from each study. Sutton et al.[18] note that these tests can give considerably contrasting conclusions, and is difficult to offer practical advice on which test to use.

4. Example to Study the Success Rate of Antifungal Onychomycosis Lacquers

In this section we use the data from Einarson[10] to illustrate the methods of this paper, where the success rate is around 0.5 (not being close to 0 or 1). Einarson[10] illustrates the meta analysis using the success rate as the effect size. The data is collected from published and unpublished clinical trials of topical antifungal lacquers used in the treatment of onychomycosis (i.e., fungal infection of the nails). Two products ciclopirox and amorolfine are of interest. The studies were identified by searching MEDLINE, EMBASE, and International Pharmaceutical Abstracts. The information about accepted trials for ciclopirox is given in Table 2. The data related to only ciclopirox (which involves 11 studies) is reported in Table 2, and the data related to amorolfine (which involves 3 studies only) is not reported in our analysis.

The success rate is defined as the proportion of patients with onychomycosis who responded to 6 months of treatment with an antifungal lacquer. The

Table 2. Studies about Ciclopirox.

Study Name	Number of Successes	Number of Patients	Success Rate
Adam	28	42	0.667
Baran *et al.*	13	25	0.520
Ehlers and Hubner	33	43	0.767
Effendy and Luders	20	22	0.909
Friederich and Effendy	55	56	0.982
Nolting and Lassus 1	28	30	0.933
Nolting and Lassus 2	33	45	0.733
Seebacher *et al.*	911	1154	0.789
Wu 1	80	100	0.800
Wu 2	77	100	0.770
Wu 3	94	102	0.922
Total	1372	1719	

result is considered as a success if the patient is cured or improved (i.e., at least moderate improvement in the patient's condition), and considered as a failure otherwise. The results of meta analysis with five possible effect sizes are considered. The estimate of p with these models are $0.85, 0.79, 0.86, 0.80$ and 0.80 respectively using the fixed effects model, and $0.82, 0.80, 0.82, 0.82$ and 0.79 respectively using the random effects model. These estimates are pretty close and any of the models give similar results. However the value of Q which depends on the transformed values of $\hat{p}$ vary greatly, the values being $119.9, 36.2, 118.6, 60.2$ and 9.4 respectively, under the five models. The p-value associated with $Q = 9.4$, corresponding to Model 5 (0.4979) does not indicate significance of τ^2, and $\hat{\tau}^2 = 0$. Model 5 which uses the variance stabilizing transformation based on Poisson assumption is not appropriate as p is not too small in this example.

Confidence intervals for I^2, τ^2, τ, p (with fixed effects model), and p (with random effects model) are also given in Table 3. The confidence intervals for p are similar with all the models. However, we suggest Model 4 which is based on the variance stabilizing transformation $T = \sin^{-1}(\sqrt{\hat{p}})$, and provides the confidence interval for p as $(0.75, 0.88)$.

Einarson[10] reported that the success rate for ciclopirox is estimated as 0.816 with s.e. 0.035 (using model 1 with random effects), and 95% confidence interval $(0.747, 0.885)$. Similarly the success rate for amorolfine is estimated as 0.714 with s.e. 0.023, with 95% confidence interval $(0.669, 0.760)$. These two intervals overlap and we may conclude that there is no significant difference between the success rates of ciclopirox and amorolfine. However, if we calculate the 95% confidence interval for the difference of the two success rates, we get $(0.816 - 0.714) \pm 1.96\sqrt{(0.035^2 + 0.023^2)} = 0.102 \pm 0.082 = (0.020, 0.184)$ i.e., the success rate for Ciclopirox is better than that of amorolfine by 2% to 18%. Einarson[10] also points out that clinical significance also need to be considered, not just statistical significance of the difference.

Table 3. Meta analysis of Ciclopirox data. Number of studies, $k = 11$.

	Model 1 $T = \hat{p}$	Model 2 $T = \ln(\hat{p}/(1-\hat{p}))$	Model 3 $T = \ln(\hat{p})$	Model 4 $T = \sin^{-1}(\sqrt{\hat{p}})$	Model 5 $T = \sqrt{\hat{p}} = \sqrt{X}$
$\hat{\theta}$	0.8477	1.3104	-0.1478	1.1120	0.8926
$s.e.(\hat{\theta})$	0.0084	0.0609	0.0098	0.0121	0.0120
$\hat{p}$	0.8477	0.7876	0.8626	0.8039	0.7968
Q	119.868	36.230	118.633	60.156	9.365
$P-value$	<0.0001	0.0001	<0.0001	<0.0001	0.4979
$\hat{\tau}^2$	0.0111	0.2020	0.0146	0.0136	0
$s.e.(\hat{\tau}^2)$	0.0087	0.1749	0.0118	0.0106	0.0012
$\hat{\theta}^*$	0.8158	1.3748	-0.1977	1.1324	0.8897
$s.e.(\hat{\theta}^*)$	0.0353	0.1793	0.0418	0.0410	0.0063
$\hat{p}^*$	0.8158	0.7981	0.8206	0.8198	0.7916
$I^2(\%)$	91.658	72.398	91.571	83.377	0
CI for τ^2	(0.0067,0.0177)	(0.0750,0.4353)	(0.0089,0.0234)	(0.0069,0.0251)	(0,0.0037)
CI for τ	(0.0824,0.1331)	(0.2738,0.6597)	(0.0945,0.1529)	(0.0830,0.1584)	(0,0.0609)
CI for $I^2(\%)$	(87.077,94.614)	(49.329,84.965)	(86.925,94.565)	(71.699,90.236)	(0,57.694)
CI for p (fem)	(0.8313,0.8641)	(0.7669,0.8069)	(0.8461,0.3794)	(0.7848,0.8223)	(0.7551,0.8395)
CI for p (rem)	(0.7465,0.8850)	(0.7356,0.8489)	(0.7561,0.3907)	(0.7542,0.8771)	(0.7697,0.8138)

A graphical method of presenting the point estimates and confidence intervals of the effect sizes for each study along with the estimate and confidence interval for the overall effect size is done through forest plots. For the Ciclopirox study, this plot is displayed in figure 1.

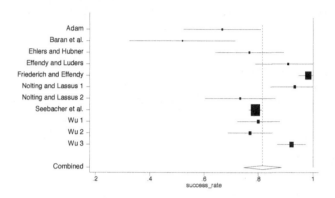

Fig. 1. Forest Plot of studies about Ciclopirox.

We note that if Models 2-5 are used, we can get confidence intervals on the difference of the effect sizes. In the next section we present an example where the success rate is very small.

5. Example to Study the Incidence of Esophageal Adenocarcinoma (EAC) in Non-dysplastic Barrett's Esophagus (NDBE)

Desai et al.[8] searched MEDLINE and EMBASE regarding studies about the incidence of EAC in NDBE. 73 studies were included with a total of 13,027 patients and 59,861 patient years follow up. The results are given in Table 5.

We briefly explain the terminology used in this study. The esophagus carries food and liquids from the mouth to the stomach. Barrett's esophagus is a condition in which the tissue lining the esophagus is replaced by tissue that is similar to the lining of the intestine. This process is called intestinal metaplasia. Barrett's esophagus can only be diagnosed using an upper gastrointestinal (GI) endoscopy to obtain biopsies of the esophagus. Because Barrett's esophagus does not cause any symptoms, many physicians recommend that adults older than 40 who have had gastroesophageal reflux disease (GERD) for a number of years undergo an endoscopy and biopsies to check for the condition. A small number of people with Barrett's esophagus develop a rare but often deadly type of cancer of the esophagus. Typically, before esophageal cancer develops, precancerous cells appear in the Barrett's tissue. This condition is called dysplasia and can be seen only through biopsies. Detecting and treating dysplasia may prevent cancer from developing.

In Table 4, the relevant information from 73 studies is provided. For each study, the author and year of publication, patient-years in follow-up (n), and the number of incident cancers X, and the incidence rate per 1000 person years $(1000x/n)$ is given in the table. In each study, the number of patients participating in the study is not provided here, as that information is not used in the analysis. But this information is given in Desai et al. (2011). When $X = 0$ in a study, then $\hat{p} = x/n = 0$ and estimate of variance of $\hat{p}$ is zero. This implies that the weights w can not be calculated unless some adjustments are made to $\hat{p}$. In the meta analysis literature, several suggestions are made for the adjustments. One widely used estimate is $\hat{p} = 0.5/(n + 0.5)$, when $x = 0$ (Hartung, Knapp and Sinha[13]). This adjustment may have a severe effect in the studies with smaller sample size, and may inflate the combined estimate of the effect size. An advantage with models using the variance stabilizing transformation is that they can be used without the adjusted ratios.

In Table 5, the results of the meta analysis are given. The incidence rates per 1000 patients years follow up for the three models using $T = \hat{p}$, $T = \ln(\hat{p}/(1 - \hat{p}))$, and $T = \ln(\hat{p})$, are 3.1, 5.3, 5.4 respectively with the fixed effects model, and 3.2, 5.2, 5.2 respectively with random effects model. Models 2 and 3 are giving very similar results as p is very close to zero. The model with $T = \sin^{-1}(\sqrt{\hat{p}})$, using adjusted rates and unadjusted rates, and model with $T = \sqrt{\hat{p}}$ using unadjusted rates are also compared. The incidence rates per 1000 patients years follow up for these three models are 3.9, 3.5 3.5 respectively with the fixed effects model, and 4.2, 3.6, 3.6 with random effects model.

With model 1, i.e., $T = \hat{p}$ has $Q = 75.3475$ and has a corresponding P-value 0.3707. This indicates that fixed effects and random effects models give similar results. However, Q values with all other models in Table 5 have much larger values ranging from 100 to 144 and very small P-values associated with them. Also, model 1 has $I^2 = 4.4\%$, and other models have ranging from 28% to 50%.

243

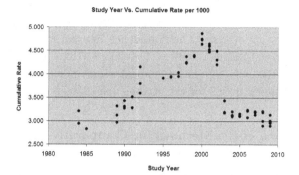

Fig. 2. Cumulative rate per 1000 by Year of study.

Cumulative meta analysis: This analysis gives results accumulated over successive studies. When the data are sorted by year, this would show the conclusions that could have been obtained at any point in time with each new study's appearance. The plot of unadjusted rate vs. year of study indicates that generally the rates are decreasing over time. The trend is particularly clear after the year 2000. From figure 2., the graph of Cumulative rate by Year of study, the following observations can be made: (i) The cumulative rates ranged from 2.8 to 4.9. (ii) The cumulative rates were steadily increasing from 1984 (2.9) to 2000 (4.9), and then steadily decreasing between 2000 (4.9) and 2009 (around 3.0). (iii) From 2002 to 2003, there is a big drop of the cumulative rate from 4.2 to 3.2, and the rate remained around 3.0 from 2003 to 2009.

A similar cumulative analysis is done by size of study (i.e., Person-years), and the following observations are made: Note that the cumulative rates are determined starting with the bigger studies first, and smaller studies at the end. (i) The cumulative rates ranged from 2.3 to 3.4. (ii) These cumulative rates are not as varied as the cumulative rates based on year of study. Starting with the biggest study (based on 4758 person-years) with a rate of 2.3, the cumulative rate increased to 3.2 (with the study based on 3711 person-years), and stayed around that rate even for smaller studies with the over all rate of 3.1.

In order to see if there is a significant change between (i) the studies upto 1999 and (ii) the studies 2000 onwards, a confidence interval for the difference between the rates is calculated. The rate per 1000 patient years of follow up for group (i) based on 26 studies is 4.400 and s.e 0.733 with 95%

confidence interval (2.964, 5.837), and for group (ii) based on 47 studies is 3.045 and s.e. 2.67 with 95% confidence interval (2.522, 3.568). These confidence intervals overlap. The confidence interval for the difference of the two rates is $(1.355 \pm 1.560) = (-0.205, 2.915)$. The difference even though not statistically significant, indicates in a practical sense, the rate might have dropped by 2.9 for the studies done 2000 onwards.

6. Modification of the Methods for Comparative Studies

Let p_1 and p_2 denote the population proportions of the two groups (experimental and control). The commonly used measures of the effect size θ are:

$$\theta_1 = p_1 - p_2$$
$$\theta_2 = \ln(\frac{p_1}{1-p_1} \frac{1-p_2}{p_2}) = \ln(\frac{p_1}{1-p_1}) - \ln(\frac{p_2}{1-p_2})$$
$$\theta_3 = \ln(\frac{p_1}{p_2}) = \ln(p_1) - \ln(p_2)$$
$$\theta_4 = \sin^{-1}(\sqrt{p_1}) - \sin^{-1}(\sqrt{p_2}).$$

These measures are such that the value 0 indicates no difference between the two groups. Suppose n_1 and n_2 are the sample sizes of the two groups, and X_1 and X_2 are the number of units with the attribute under study, then the estimates of p_1 and p_2 are given by $\hat{p}_1 = X_1/n_1$ and $\hat{p}_2 = X_2/n_2$ respectively. The estimates of θ are calculated as

$$\hat{\theta}_1 = \hat{p}_1 - \hat{p}_2$$
$$\hat{\theta}_2 = \ln(\frac{\hat{p}_1}{1-\hat{p}_1} \frac{1-\hat{p}_2}{\hat{p}_2}) = \ln(\frac{\hat{p}_1}{1-\hat{p}_1}) - \ln(\frac{\hat{p}_2}{1-\hat{p}_2})$$
$$\hat{\theta}_3 = \ln(\frac{\hat{p}_1}{\hat{p}_2}) = \ln(\hat{p}_1) - \ln(\hat{p}_2)$$
$$\hat{\theta}_4 = \sin^{-1}(\sqrt{\hat{p}_1}) - \sin^{-1}(\sqrt{\hat{p}_2}),$$

and the corresponding variances of the estimates are:

$$Var(\hat{\theta}_1) = \frac{p_1(1-p_1)}{n_1} + \frac{p_2(1-p_2)}{n_2},$$
$$Var(\hat{\theta}_2) = \frac{1}{n_1 p_1(1-p_1)} + \frac{1}{n_2 p_2(1-p_2)},$$
$$Var(\hat{\theta}_3) = \frac{1-p_1}{n_1 p_1} + \frac{1-p_2}{n_2 p_2},$$
$$Var(\hat{\theta}_4) = \frac{1}{4n_1} + \frac{1}{4n_2}.$$

If $X_1 = 0$ and/or $X_2 = 0$, note that $Var(\hat{\theta}_1) = 0$, and its reciprocal to determine the weight $w = 1/Var(\hat{\theta}_1)$ can not be calculated, and also $Var(\hat{\theta}_2)$, $Var(\hat{\theta}_3)$ can not be calculated. The adjusted values need to be used. However, the variance stabilizing transformation does not need the adjustment.

Table 4. Incidence Rates($\hat{p}$) and Adjusted Incidence Rates(adj $\hat{p}$) per 1000 person-years.

No.	Study Name	No. of incident Cancers (X)	Patient years in follow-up (n)	$1000 * \hat{p}$	$1000 * \widehat{adj\,p}$
1	Sharma(2006)	10	2315	4.32	4.32
2	Dulai(2003)	2	1618	1.24	1.24
3	Gladman(2006)	4	1068	3.75	3.75
4	Hage(2004)	3	812	3.69	3.69
5	Basu(2004)	0	383	0.00	1.30
6	Murray(2003)	11	4471	2.46	2.46
7	Miros(1991)	0	249	0.00	2.00
8	Conio(2003)	3	825	3.64	3.64
9	Atwood(1992)	2	135	14.81	14.81
10	Williamson(1991)	4	491	8.15	8.15
11	Srinivasan(2001)	0	22	0.00	22.22
12	Wilkinson(1999)	0	115	0.00	4.33
13	Younes(1997)	0	120	0.00	4.15
14	Katz(1997)	3	563	5.33	5.33
15	Vieth(2006)	11	4758	2.31	2.31
16	Martinek(2008)	2	700	2.86	2.86
17	Cooper(2005)	3	959	3.13	3.13
18	Bani-Hani(2000)	12	1381	8.69	8.69
19	Macdonald(2000)	3	607	4.94	4.94
20	Spechler(1984)	1	340	2.94	2.94
21	Savary(1984)	5	1528	3.27	3.27
22	Watson(1990)	0	108	0.00	4.61
23	Rana(2001)	1	285	3.51	3.51
24	Cameron(1985)	2	884	2.26	2.26
25	Eckhardt(2001)	2	600	3.33	3.33
26	Iftikhar(1992)	4	462	8.66	8.66
27	Csendes2004	5	1795	2.79	2.79
28	Parilla(2002)	3	600	5.00	5.00
29	Ovaska(1989)	3	166	18.07	18.07
30	Moghissi(1992)	4	298	13.42	13.42
31	Streitz(1998)	7	510	13.73	13.73
32	Murphy(2005)	1	380	2.63	2.63
33	Oberg(2005)	3	812	3.69	3.69
34	Hameetman(1989)	2	223	8.97	8.97
35	McDonald(1996)	3	728	4.12	4.12
36	Conio(2001)	4	585	6.84	6.84
37	Switzer(2008)	5	687	7.28	7.28
38	Reid(2000)	5	503	9.94	9.94
39	Hillman(2008)	3	2309	1.30	1.30
40	Skinner(1989)	3	145	20.69	20.69
41	Meining(2004)	0	40	0.00	12.35
42	Lim(2007)	11	2576	4.27	4.27
43	Spechler(2001)	4	1037	3.86	3.86
44	Kelty(2007)	17	4548	3.74	3.74
45	klinkenberg(2000)	1	456	2.19	2.19
46	peters (1999)	0	106	0.00	4.69
47	malesci (1996)	0	14	0.00	34.48
48	Bowers(2002)	0	337	0.00	1.48
49	Low(1999)	0	15	0.00	32.26
50	Chen (2001)	0	148	0.00	3.37
51	Demeester(1990)	0	72	0.00	6.90
52	Lundell(2009)	1	288	3.47	3.47
53	Patti(1999)	0	32	0.00	15.38
54	Sagar(1995)	1	308	3.25	3.25
55	Williamson (1990)	3	151	19.87	19.87
56	Luostarinen(1998)	0	33	0.00	14.93
57	Nguyen(2009)	13	2620	4.96	4.96
58	Alcedo(2009)	7	1358	5.15	5.15
59	Bright(2009)	4	776	5.15	5.15
60	Olithselvan(2007)	3	425	7.06	7.06
61	csendes2009	0	435	0.00	1.15
62	Vonrahden(2008)	5	2738	1.83	1.83
63	O'Riordan(2004)	0	184	0.00	2.71
64	Gurski(2002)	0	277	0.00	1.80
65	Oellschlager(2003)	0	187	0.00	2.67
66	Zaninotto(2005)	0	77	0.00	6.45
67	Biertho(2007)	0	281	0.00	1.78
68	Gatenby(2009)	20	3711	5.39	5.39
69	Desai(2001)	0	155	0.00	3.22
70	Abbas(2004)	0	100	0.00	4.98
71	Aldulaimi(2005)	12	338	35.50	35.50
72	Fitzgerald(2001)	2	483	4.14	4.14
73	Teodori(1998)	4	390	10.26	10.26

Table 5. Meta analysis of Cancer Incidence data. Number of Studies, $k = 73$.

	Model 1 $T = \hat{p}$	Model 2 $T = \ln(\hat{p}/(1-\hat{p}))$	Model 3 $T = \ln(\hat{p})$	Model 4 $T = \sin^{-1}(\sqrt{\hat{p}})$	Model 4 $T = \sin^{-1}(\sqrt{\hat{p}})$ with unadjusted data	Model 5 $T = \sqrt{\hat{p}} = \sqrt{\bar{X}}$ with unadjusted data
$\hat{\theta}$	0.0031	-5.2277	-5.2250	0.0623	0.0593	0.0592
$s.e(\hat{\theta})$	0.0002	0.0624	0.0620	0.0020	0.0020	0.0020
$\hat{p}$	0.0031	0.0053	0.0054	0.0039	0.0035	0.0035
Q	75.348	136.346	138.542	100.457	143.706	142.826
$P - value$	0.3707	<0.0001	<0.0001	0.0150	<0.0001	<0.0001
$\hat{\tau}^2$	0.0000002	0.258962	0.263872	0.000123	0.000309	0.000305
$s.e(\hat{\tau}^2)$	0.0000007	0.103673	0.104217	0.000077	0.000123	0.000122
$\hat{\theta}^*$	0.0032	-5.2482	-5.2513	0.0650	0.0599	0.0599
$s.e(\hat{\theta}^*)$	0.0002	0.0981	0.0982	0.0027	0.0034	0.0034
$\hat{p}^*$	0.0032	0.0052	0.0052	0.0042	0.0036	0.0036
$I^2(\%)$	4.443	47.193	48.030	28.327	49.898	49.589
CI for τ^2	(0,0.000001)	(0.1271,0.4325)	(0.1323,0.4358)	(0.000012,0.000271)	(0.000162,0.000502)	(0.000159,0.000497)
CI for τ	(0,0.0012)	(0.3566,0.6576)	(0.3638,0.6609)	(0.00035,0.0165)	(0.0127,0.0224)	(0.0126,0.0223)
CI for $I^2(\%)$	(0,27.817)	(30.495,59.879)	(31.671,60.473)	(3.776,46.614)	(34.290,61.798)	(33.858,61.579)
CI for p (fem)	(0.0026,0.0035)	(0.0047,0.0060)	(0.0048,0.0061)	(0.0034,0.0044)	(0.0031,0.0040)	(0.0031,0.0040)
CI for p (rem)	(0.0027,0.0036)	(0.0043,0.0063)	(0.0043,0.0063)	(0.0036,0.0049)	(0.0028,0.0044)	(0.0028,0.0044)

References

1. C. B. Begg and M. Mazumdar, *Biometrics* **50**, 1088 (1994).
2. M. Borenstein, L. V. Hedges, J. P. T. Higgins and H. R. Rothstein, *Introduction to Meta-Analysis* (John Wiley & Sons Ltd., Chichester, UK, 2009).
3. G. Casella and R. L. Berger, *Statistical Inference*, 2nd edition. (Duxbury Press, Pacific Grove, CA, 2002).
4. H. Chen, *Journal of American Statistical Association* **85**, 514 (1990).
5. C. J. Clopper and E. S. Pearson, *Biometrika* **26**, 404 (1934).
6. W. G. Cochran, *Journal of the Royal Statistical Society* Supplement **4**, 102 (1937).
7. R. DerSimonian and N. Laird, *Controlled Clinical Trials* **7**, 177 (1986).
8. T. Desai, K. Krishnan, N. Samala, J. Singh, J. Cluley, P. Subbaiah and C. Howden, The incidence of esophageal adenocarcinoma in non-dysplastic Barrett's esophagus: A meta-analysis. *Submitted*, (2011).
9. M. Egger, G. D. Smith, M. Schneider and C. Minder, *British Medical Journal* **315**, 629 (1997).
10. T. R. Einarson, *Clinical Therapeutics* **19**, 559 (1997).
11. D. A. Follmann and M. A. Proschan, *Biometrics* **55**, 732 (1999).
12. J. Hartung and G. Knapp, *Statistics in Medicine* **20**, 3875 (2001).
13. J. Hartung, G. Knapp and B. K. Sinha, *Statistical Meta-Analysis with Applications* (John Wiley & Sons Inc., Hoboken, NJ, 2008).
14. L. V. Hedges and I. Olkin, *Statistical Methods for Meta-Analysis* (Academic Press, Boston, 1985).
15. J. Higgins, S. G. Thompson, J. J. Deeks and D. G. Altman, *British Medical Journal* **327**, 557 (2003).
16. D. Jackson, J. Bowden and R. Baker, *Journal of Statistical Planning and Inference* **140**, 961 (2010).
17. G. Knapp, B. J. Biggerstaff and J. Hartung, *Biometrical Journal* **48**, 271 (2006).
18. A. J. Sutton, K. R. Abrams, D. R. Jones, T. A. Sheldon and F. Song, *Methods for Meta-Analysis in Medical Research* (John Wiley & Sons Inc., Chichester, UK, 2000).
19. A. J. Sutton and J. P. T. Higgins, *Statistics in Medicine* **27**, 625 (2008).

RISK REDUCTION OF THE SUPPLY CHAIN THROUGH POOLING LOSSES IN CASE OF BANKRUPTCY OF SUPPLIERS USING THE BLACK-SCHOLES-MERTON PRICING MODEL

RAUL VALVERDE* and MALLESWARA TALLA

*Department of Decision Sciences and MIS, Concordia University,
1455 de Maisonneuve Blvd West, Montreal, Quebec, Canada*
* E-mail: rvalverde@jmsb.concordia.ca

In recession times, slower demand, shrunk liquidity, and increasing pressure on cost can lead to bankruptcy of suppliers. The risks due to supplier bankruptcy include (a) losses due to supply chain disruption, (b) delayed or stopped finished goods shipments, (c) difficulty in finding cost-effective alternate suppliers and sourcing contracts, (d) emergency procurements, (e) loss of reputation and market share loss, etc. Bankruptcy models can be used to estimate the probability that a supplier may go bankruptcy, and a level of probability can be established that triggers the risks. This paper uses the Black-Scholes-Merton option pricing model for estimating the probability of bankruptcy of supplier by extracting and examining the riskiness in stock market price of supplier. The paper uses the pooling arrangements among companies that source from multiple suppliers as a way to reduce the risk due to supplier bankruptcy.

1. Introduction

In recent times, companies are increasingly forming global supply chains and favoring global sourcing practices for lowering the purchase prices. While the global sourcing truly offered the expected benefits in the short run, it increased the risk of facing several challenges in the long run. One of the major issues is the supplier financial distress leading to supplier bankruptcy due to slower demand, shrunk liquidity, and increasing pressure on cost. The risks due to supplier bankruptcy include (a) losses due to supply chain disruption, (b) delayed or stopped finished goods shipments, (c) difficulty in finding cost-effective alternate suppliers and sourcing contracts, (d) emergency procurements, (e) loss of reputation and market share loss, etc. In summary, if a supplier becomes bankrupt, that firm may not be able to meet its entire customer requirements in the short-term, and will not meet any customer requirements if it eventually goes out of business (see Zsidisin and Wagner[7]). A firm is obliged to evaluate the financial viability of suppliers in order to avoid the consequences of supplier default, insolvency, or bankruptcy (see Milne[3] and Wagner and Johnson[6]).

The good tools for financial evaluation are the bankruptcy models, as these can be used for estimating the probability that a supplier may go bankrupt, and a level of probability can be established that triggers the risks. This paper uses the Black-Scholes-Merton option pricing model[1] for estimating the probability of bankruptcy of supplier by extracting and examining the riskiness in stock market price of supplier. The model assumes: maturity of liabilities equals one year; the dividend rate is based on the sum of common dividends, preferred dividends, and interest expense; call option equation has been modified to account for the fact that shareholders receive common dividends.[1]

In order to minimize the risk of bankruptcy among suppliers, the paper proposes the use of a pooling arrangement among companies. In a pooling arrangement, every participant agrees to share losses equally, each paying an average loss. The arrangement does not change the expected loss but reduces uncertainty because the variance decreases. This makes the losses due to supplier bankruptcy become more predictable since the maximum probable loss declines and the distribution of costs becomes more symmetric. The predictability increases with the number of participants and decreases with correlation in losses. Previously, pooling arrangements have been used in supply chains for inventory management for the reduction of demand variability that can lead to a reduction of safety stock and average inventory,[4] however, this paper combines the use of bankruptcy prediction models and pool arrangements for risk reduction in the supply chain. This pool arrangement methodology could be used as a tool for the preparation of insurance policies that can be sold to companies in order to protect them against supplier bankruptcy.

2. Black-Scholes-Merton Option Pricing Model

The bankruptcy losses for suppliers are based on the Black-Scholes-Merton option pricing model1. With this model, a company is defined as being in bankruptcy if its corporate value shown by aggregate market value falls short of its amount of debt.[1] The model captures the likelihood that the values of firms assets will decline to such an extent that the firm will be unable to repay its debts.[1]

Equity can be viewed as a call option on the value of the firm's assets.

The strike price of the call option is equal to the face value of the firm's liabilities and the option expires at time T when the debt matures 1.

The equation for valuing equity as a call option on the value of the firm's assets is given in Eq. (1) below. This equation is modified for dividends and reflects that the stream of dividends paid by the firm accrue to the equity holders.

The BSM equation is given by (see Wagner and Johnson[6]):

$$V_E = V_A e^{-\delta T} N(d_1) - X e^{-rT} N(d_2) + (1 - e^{-\delta T}) V_A \qquad (1)$$

where $N(.)$ represents the cumulative distribution function of standard normal random variable and d1 and d2 are given by:

$$d_1 = \frac{\ln\left(\frac{V_A}{X}\right) + \left(r - \delta + \frac{\sigma_A^2}{2}\right) T}{\sigma_A \sqrt{T}} \qquad (2)$$

$$d_2 = d_1 - \sigma_A \sqrt{T} = \frac{\ln\left(\frac{V_A}{X}\right) + \left(r - \delta - \frac{\sigma_A^2}{2}\right) T}{\sigma_A \sqrt{T}} \qquad (3)$$

where $V_E \equiv V_E(t)$ is the current market value of equity, $V_A \equiv V_A(t)$ is the current (at time t) market value of assets, X is the face value of debt maturing at time T, r is the continuously-compounded risk-free rate, δ is the continuous dividend rate σ_A is the standard deviation of asset returns.

Under the BSM model, the probability of bankruptcy is simply the probability that the market value of assets, V_A is less than the face value of the liabilities, X, at time T. The BSM model assumes that the natural log of future asset values is normally distributed. The probability of bankruptcy is a function of the distance between the current value of the firm's assets and the face value of its liabilities, adjusted for the expected growth in asset values relative to asset volatility.

As shown in McDonald,[2] the probability of bankruptcy can be calculated as:

$$N\left(\frac{\ln\left(\frac{V_A}{X}\right) + \left(\mu - \delta + \frac{\sigma_A^2}{2}\right) T}{\sigma_A \sqrt{T}}\right) \qquad (4)$$

where μ is the continuously-compounded expected return on assets. Hillegeist et al.[5] provided the SAS code for the estimation of V_A and σ_A by simultaneously solving the BSM equation (1) and the optimal hedge equation given by

$$\sigma_E = \frac{V_A \mathrm{e}^{-\delta t} N(d_1)\sigma_A}{V_E} \tag{5}$$

through iterative process. These values are then used to estimate μ from the following formula

$$\mu = \max\left(\frac{V_A(t) - V_A(t-1) + dividents}{V_A(t-1)}, r\right) \tag{6}$$

where dividends in the above equation represent the sum of the common and preferred dividends declared during the year. This in turn can finally be used to calculate the probability of bankruptcy.

The basic idea for estimating the probability of a supplier company bankruptcy is to recognize the stock price movement pattern of the supplier company, and evaluate the historic events information, which is available to public via company press meets, market focus, etc. The procedure for extracting such information was developed by Hillegeist et al.[5] This paper describes the reasoning behind using stock prices, as opposed to accounting data, to extract the probability of supplier bankruptcy estimates as well as the methodological steps and assumptions behind the estimates.

According to option-pricing theories (see McDonald[2]), a market-based measure, that is called Black-Sholes-Model probability of bankruptcy (BSM-PB), should use all available information about the probability of bankruptcy.[5] The BSM-PB contains relatively more information than just the Score variables used traditionally for bankruptcy prediction, however the accounting measures will not be incrementally informative to BSM-PB. Hillegeist et al.[5] tested the validity of these implications using a large sample consisting of 65,960 firm-year observations including 516 bankruptcies during the 1979-1997 period. They found that BSM-PB has relatively more explanatory power than either of the two Scores, even when the Scores are decomposed to reflect industry differences or annual changes.

The model assumes that volatility is a crucial variable in bankruptcy prediction since it captures the likelihood that the values of firms assets will decline to such an extent that the firm will be unable to repay its debts. Equity can be viewed as a call option on the value of the firm's assets. The strike price of the call option is equal to the face value of the firm's liabilities and the option expires at time T when the debt matures.

3. Risk Reduction through Pooling Independent Losses

A risk pool is one of the forms of risk management practiced in insurance. Pooling arrangements do not change a company's expected loss, but reduce the uncertainty (standard deviation) of a loss. Risk pooling arrangements make each participant's loss more predictable (Zsidisin and Wagner[7]).

Correlation analysis is very important in pooling arrangements. A positive correlation in losses is less desirable than null correlation (uncorrelated losses) in the context of risk management. While a positive correlation in losses reduces the extent to which risk pooling lowers the standard deviation of losses, null correlation in losses increase it. The concept of pooling losses has been used in supply chain (see Milne[3]). Risk pooling suggests that demand variability is reduced if one aggregates demand across locations because as demand is aggregated across different locations, it becomes more likely that high demand from one customer will be offset by low demand from another. This reduction in variability allows a decrease in safety stock and therefore reduces average inventory, this suggests that the use centralized warehouses would be able to reduce inventory costs as it reduces safety stock but this benefit will decrease as the correlation between demands demanding inventory becomes positive (see Milne[3]).

The expected loss and variance for a company C associated with M suppliers, due to supplier bankruptcy can be estimated by

$$\bar{L}_C = \sum_{i=1}^{M} L_i P_i, \tag{7}$$

and

$$\sigma_C^2 = \sum_{i=1}^{M} (L_i - \bar{L}_C)^2 P_i, \tag{8}$$

respectively, where L_i is the loss of a company due to supplier bankruptcy, P_i is the probability of bankruptcy of a supplier.

In the context of the proposed research, each firm in the pool is willing to share the losses generated due to bankruptcy of suppliers. The research will show how the mechanism can be used to reduce the standard deviation of the losses associated with risk of losses.

4. Illustration of the Methods on Real Data

In order to find the probability of bankruptcy, a research was conducted by selecting 23 companies that trade in the stock market and publish their financial statements. The BSM model has been used for calculating the probability of bankruptcy for these 23 firms for the first quarter for the year 2000. A scenario of two companies that have three suppliers each was generated and the losses due to bankruptcy for each supplier were pooled among the two companies and the expected loss and standard deviation of losses calculated by pooling losses. The objective of the research is to show that the risk pooling contract minimizes the risk exposure.

A SAS program generated by Hillegeist *et al.*[5] was used to calculate the probability of bankruptcy. The calculation was performed in three steps. In the first step, the values of V_A and σ_A were estimated by simultaneously solving the call option B-S-M equation (Eq. (1)) and the optimal hedging equation (Eq. (4)).

In the initial step, V_E was set equal to the total market value of equity based on the closing price at the end of the firm's fiscal year, σ_E was computed using daily return data from the Center for Research in Security Prices database (http://www.crsp.com) over the entire fiscal year. The strike price X was set equal to the book value of total liabilities, T was taken to be one year, and r was set at the one-year treasury bill rate. The dividend rate, δ, was the sum of the prior year's common and preferred dividends divided by the approximate market value of assets, which is defined as total liabilities plus the market value of equity.

In the second step, the expected market return on assets, μ, was calculated based on the actual return on assets during the previous year and with the help of Eq. (6). This process is based on the estimates of VA that were computed in the previous step.

Finally, the values for $V_A, \sigma_A, \delta, T, X$ and μ were used to calculate the probability of bankruptcy for each firm-year via Eq. (4). To do this, the value inside the parentheses in Eq. (4) was first calculated to then determine the probability of bankruptcy corresponding to this value using the standard normal distribution.

Table 1 presents the probability of bankruptcy for the 23 companies based on the first quarter of the year 2000. The same table displays the values of V_A and σ_A calculated in the first step.

The scenario used for this analysis assumes two companies with three suppliers each. The loss due to bankruptcy for each supplier is assumed to be constant and estimated to be $5000, which include losses due to supply chain disruption, delayed or stopped finished goods shipments, difficulty in finding cost-effective alternate suppliers and sourcing contracts, emergency procurements, loss of reputation and market share among the possi-

Table 1.   Probabilities of bankruptcy for suppliers for Year 2000 and Quarter 1.

Supplier company	GVKey	V_A	σ_A	Probability of bankruptcy
AAR CORP	1004	708.5085	0.263461	.02170124
ADC TELECOMMUNICATIONS INC	1013	10381.02	0.645216	9.22E-08
ALPHARMA INC	1034	1522.568	0.237037	0.00142876
UNITED DOMONION INDUSTRIES	1036	2076.659	0.118494	0.0002026
AMC ENTERTAINMENT INC	1038	1167.491	0.066569	0.10188689
AMR CORP/DE	1045	21949.29	0.094827	0.00510627
CECO ENVIRONMENTAL CORP	1050	58.37434	0.421999	0.22111509
ASA BERMUDA LTD	1062	176.6768	0.385425	0
AVX CORP	1072	4321.04	0.651542	0.00019655
PINNACLE WEST CAPITAL	1075	6719.818	0.099677	7.56E-06
AARON RENTS INC	1076	373.1073	0.285872	.00019291
ABITIBI CONSOLIDATED INC	1081	3063.677	0.06287	.04249133
ABRAMS INDUSTRIES INC	1082	97.4321	0.117798	0.04888238
ACKERLY GROUP INC	1095	790.4289	0.196604	0.00072447
ACMAT CORP	1097	101.7899	0.203711	0.18243195
ACME UNITED CORP	1104	22.51936	0.452804	0.07257192
ACTION PRODUCTS INTL INC	1109	6.425416	0.946431	0.20192717
ACTIVISION INC	1111	307.8869	0.378032	0.0435724
RELM WIRELESS CORP	1117	36.01124	1.057167	0.29516477
ADAMS RESOURCES & ENERGY INC	1121	342.5629	0.066581	1.50E-06
AERO SYSTEMS ENGINEERING INC	1154	22.2858	0.94807	0.44375006
ADVANCES MICRO DEVICES	1161	11023.84	0.559808	0.00014882
ASM INTERNATIONAL NV	1166	1592.034	0.733109	0.03627255

Table 2.   Company A's expected losses and standard deviation due to bankruptcy of suppliers.

Suppliers	Probability of of bankruptcy	Loss due to bankruptcy	Supplier's expected loss
AAR CORP	0.021701241	5000	108.51
ABRAMS INDUSTRIES INC	0.048882378	5000	244.41
ACTION PRODUCTS INTL INC	0.201927167	5000	1009.64
Expected Loss			1362.55
Standard deviation			1898.84

ble losses. Tables 2 and 3 present the expected losses and standard deviation of companies A and B due to bankruptcy of their three suppliers, respectively. Table 4 shows the expected losses for the pooling arrangement of the two companies. In this arrangement, losses are shared equally between companies A and B. The estimates show that uncertainty represented by the standard deviation is reduced while the average loss $((1362.55+1479.45)/2 = \$ 1421)$ remained unchanged.

Table 3. Company B's expected losses and standard deviation due to bankruptcy of suppliers.

Suppliers	Probability of of bankruptcy	Loss due to bankruptcy	Supplier's expected loss
ASA BERMUDA LTD	0	5000	0
ACKERLY GROUP INC	0.00072447	5000	3.62
RELM WIRELESS CORP	0.29516477	5000	1475.82
Expected Loss			1479.45
Standard deviation			1915.03

Table 4. Pooling arrangement between companies A and B.

Losses	Portion of pooled outcome	Probability of bankruptcy	Expected loss per arrangement
0	0	0.523023	0
5000	2500	0.219406	548.51593
10000	5000	1.58793E-04	0.79396
15000	7500	0	0
5000	2500	0.17081	427.04308
10000	5000	0.07165	358.28597
15000	7500	5.1861E-05	0.38896
20000	10000	0	0
10000	5000	0.01033	51.66552
15000	7500	0.00433	32.51024
20000	10000	3.13719E-06	0.03137
25000	12500	0	0
15000	7500	1.50871E-04	1.13152
20000	10000	6.32896E-05	0.63290
25000	12500	4.58051E-08	0.00057
30000	15000	0	0
		Expected Loss	1421.00
		Standard deviation	1601.44

5. Conclusions and Future Research

This study clearly demonstrates the usefulness of estimating the probability of bankruptcy of suppliers for managing the risk to a supply chain. The results presented show how the pooling contracts can help companies to minimize the risk of losses due to supplier bankruptcy. These pooling contracts can be managed by insurance carriers and sold to companies as supplier bankruptcy insurance. Furthermore, the pooling contracts help companies minimize the risk. Here we used a simple scenario with a pooling arrangement with two companies and three suppliers. Although this paper assumes that a company does not experience any loss while a supplier is not bankrupt, unexpected events (fluctuations in price, etc.) may amount to a certain loss. The analysis and computation of this paper can be revised to account for such factors. Moreover, the future research can include different types of risks and estimates for other types of supply chain risks, and conduct a study with multiple scenarios in order to provide more evidence that the concept can work in industry. Future work can also include a software program which can be developed for any number of supplier companies and pooling contracts with increased participation.

References

1. F. Black and M. Scholes, *The Journal of Political Economy* **81**, 637 (1973).
2. R. McDonald, *Derivative Markets* (Addison Wesley, Boston, MA, 2002).
3. Richard Milne, Early Warnings in the Supply Chain, In *Financial Times Europe*, **36957**, 10 (2010).
4. D. Simchi-Levi, P. Kaminsky and E. Simchi-Levi, *Designing & Managing the Supply Chain*, second edition (McGraw Hill, New York, USA, 2003).
5. Stephen A. Hillegeist, Elizabeth K. Keating, Donald P. Cram, Kyle G. Lundstedt, *Review of Accounting Studies* **9**, 5 (2004).
6. S. M. Wagner and J. L. Johnson, *Industrial Marketing Management* **33**, 717 (2004).
7. G. A. Zsidisin and S. M. Wagner, *Journal of Business Logistics* **31**, 1 (2010).

RANDOM EFFECTS MODELLING OF NON-GAUSSIAN TIME SERIES RESPONSES WITH COVARIATES

*GUOHUA YAN, M. †TARIQUL HASAN and ‡RENJUN MA

Department of Mathematics & Statistics, University of New Brunswick, Fredericton,
NB E3B 5A3, Canada
E-mail: *gyan@unb.ca,† thasan@unb.ca, ‡renjun@unb.ca

We introduce a new class of random effects models for non-Gaussian time series data with covariates based on Tweedie distributions. Temporal dependence is accommodated by serially correlated, distribution-free random effects. This modeling approach allows us to unify the marginal and conditional inferences. An optimal estimation procedure of our models is developed using the orthodox best linear unbiased predictors of random effects. The proposed method is illustrated with two environmental examples. The first example investigates association between daily counts of emergency room visits and air pollution. The second is concerned of temporal pattern of monitored ozone levels.

1. Introduction

Time series with non-Gaussian responses frequently occur in health, biological, social and environmental sciences (see Diggle *et al.*,[3] Fahrmeir and Tutz,[6] Kedem[12] and MacDonald and Zucchini.[16] To analyze non-Gaussian time series responses with covariates, observation-driven and parameter-driven conditional modelling techniques are commonly used in literature. To accommodate temporal dependence among the time series responses under the observation-driven modelling approaches, the conditional mean of the observation at the tth time point is related to the previous $(t-1)$ observations (see Zeger and Qaqish,[20] Li[13] and Davis *et al.*[2] In the parameter-driven model, the temporal dependence is generally incorporated through latent process (see Zeger[19]). However these conditional modelling approaches fail to connect the marginal mean of the responses and the covariates directly (see Chan and Ledolter,[1] Durbin and Koopman,[5] and Jung and Liesenfeld[8]). As an alternative, marginalized modelling approach for time series responses, which relates the marginal mean and covariates directly, can be obtained by modifying the approach proposed by Heagerty and Zeger.[7] In marginalized modelling approaches, numerous artificial intermediate parameters need to be introduced that are very difficult to interpret. Therefore it is desirable to introduce a model which can unify conditional and marginal modelling interpretations.

This work is motivated by two environmental examples. In the first example, the daily counts of patients' emergency room visits due to asthma to a hospital in Prince George, British Columbia for two years were recorded along with some air pollution measurements. The second example consists of right skewed continuous observations recording the weekly maximum ozone levels in the air of Los Angeles county during a 10 years period along with the temperature and relative humidity measurements. Our scientific interest here is to model the count and right skewed continuous time series responses while accounting for a possible temporal correlation amongst the observations.

In this paper, we incorporate distribution-free random effects into Tweedie generalized linear models (TGLM) to model time series responses. This random effects TGLM is flexible to model various continuous (gamma, inverse Gaussian) and discrete (counts) responses and only uses the first two moments of the unobserved random effects. As serial dependence between observations is likely to decrease over time, we consider an auto regressive correlation structure to model the serial dependence among the time series responses. The proposed approach can also capture possible additional over dispersion in the responses. To estimate the model parameters, we introduce an optimal estimating function based on the orthodox best linear unbiased predictors (BLUP) of the random effects. Our proposed methodology is simpler and computationally more efficient as it uses the explicit form of estimating equations for regression and random effects parameters (see Ma et al.[14]). Our model also unifies conditional and marginal modelling interpretations.

After introducing the random effects Tweedie generalized linear models for time series data in Section 2, we develop the BLUP of random effects in Section 3. The estimation technique of the model parameters are presented in Section 4. The analyses of emergency room visit data and Los Angeles ozone level data are presented respectively in Section 5.

2. Mixed Model for Non-Gaussian Time Series Responses

2.1. *The model*

Let Y_t represent the non-Gaussian time series response recorded at the tth $(t = 1, 2, \ldots, T)$ time point. Then the response vector can be expressed as $\boldsymbol{Y} = (Y_1, \ldots, Y_t, \ldots, Y_T)'$. We consider the time-specific distribution-free random effect U_t for the response of the tth time point to accommodate temporal correlation among the responses. Let $\boldsymbol{U}$ denote the vector of the random effects where $\boldsymbol{U} = (U_1, \ldots, U_t, \ldots, U_T)'$. Our model is based on the following two assumptions:

Assumption 1. Time-specific random effects $U_1, \ldots, U_t, \ldots, U_T$ are serially dependent and identically distributed as

$$E(U_t) = 1, \quad \text{var}(U_t) = \tau^2 \quad \text{and} \quad \text{corr}(U_t, U_{t'}) = \rho^{|t-t'|}, \text{ for } t \neq t'.$$

This correlation structure of the random effects is known as autoregressive of order 1 (AR(1)), which is common for time series responses.

Assumption 2. Given the vector of random effects $\boldsymbol{U}$, the components of $\boldsymbol{Y}$ are conditionally independent, and the conditional distribution of Y_t given $\boldsymbol{U}$, depends on U_t only, which is

$$Y_t \mid \boldsymbol{U} \sim \text{Tw}_q(\mu_t U_t, \epsilon^2 U_t^{1-q}) \tag{1}$$

where $\mu_t = \exp(\boldsymbol{x}_t'\boldsymbol{\beta})$ with vector of covariates $\boldsymbol{x}_t$ and regression parameter vector $\boldsymbol{\beta}$. Jørgensen[9] studied a class of exponential dispersion models in which the variance of a distribution can be expressed as a dispersion parameter, ϵ^2, multiplied by a "variance function" of the mean μ, $V(\mu) = \mu^q$. Following Ma and Jørgensen,[15] the Tweedie exponential family, $\text{Tw}_q(\mu, \epsilon^2)$, with index parameter q has the form

$$f_q\left(y; \mu, \epsilon^2\right) = \begin{cases} c_q(y; \epsilon^2) \exp\left\{\frac{1}{\epsilon^2}\left(\frac{y\mu^{1-q}}{1-q} - \frac{\mu^{2-q}}{2-q}\right)\right\} & \text{if } q \neq 1, 2, \\ c_2(y; \epsilon^2) \exp\left[-\frac{1}{\epsilon^2}\left\{\frac{y}{\mu} + \log(\mu)\right\}\right] & \text{if } q = 2, \\ c_1(y) \exp\left[y\log(\mu) - \mu\right] & \text{if } q = 1, \epsilon^2 = 1, \end{cases}$$

where the index parameter q ranges in $(-\infty, 0] \cup [1, \infty)$, and the explicit expressions for $c_q(y; \epsilon^2)$ are given by Jørgensen.[11] The family unifies and generalizes a few standard distributions: normal, Poisson, gamma and inverse-Gaussian for $q = 0, 1(\epsilon^2 = 1), 2, 3$ respectively. The expression $c_q(y; \epsilon^2)$ is immaterial in our derivation of the moment structures in the following sections. In (1), the conditional mean and variance of this Tweedie family is given by $E(Y_t|U) = \mu_t U_t$ and $\text{var}(Y_t|U) = \epsilon^2 U_t^{1-q} \mu_t^q U_t^q = \epsilon^2 \mu_t^q U_t$. In this modelling approach, we introduce random effect U to capture serial dependence and additional over-dispersion of the time series responses.

2.2. Moment structure

The main focus of this section is to present the moment structures of the mixed model for non-Gaussian time series responses discussed in the previous subsection. These moments can be obtained after some algebraic calculations by methods of conditioning on time-specific random effects. The unconditional mean, variance and covariance are presented here to facilitate the parameter estimation in next section.

The unconditional expectation of the response Y_t can be expressed as

$$E(Y_t) = EE(Y_t \mid U) = \mu_t E(U_t) = \mu_t. \tag{2}$$

In this approach the log conditional mean can be expressed as $\log E(Y_t|U) = x_t'\beta + \log U_t$ and the log of marginal mean can be expressed as $\log E(Y_t) = x_t'\beta$. Our model can achieve both conditional and marginal modelling interpretations as $\log EE(Y_t|U) = \log E(Y_t) = x_t'\beta$.

The unconditional variance of the responses Y_t has the form

$$\begin{aligned}
\text{var}(Y_t) &= E\{\text{var}(Y_t \mid U)\} + \text{var}\{E(Y_t \mid U)\} \\
&= \epsilon^2 \mu_t^q E(U_t) + \mu_t^2 \text{var}(U_t) = \epsilon^2 \mu_t^q + \tau^2 \mu_t^2.
\end{aligned} \tag{3}$$

Similarly the unconditional covariance of the responses Y_t and $Y_{t'}$ is

$$\begin{aligned}
\text{cov}(Y_t, Y_{t'}) &= E\{\text{cov}(Y_t, Y_{t'} \mid U)\} + \text{cov}\{E(Y_t \mid U), E(Y_{t'} \mid U)\} \\
&= \mu_t \mu_{t'} \text{cov}(U_t, U_{t'}) = \mu_t \mu_{t'} \tau^2 \rho^{|t-t'|}.
\end{aligned} \tag{4}$$

In next section we present the best linear unbiased predictors (BLUP) of the random effects, which will be used to estimate the regression parameters and random effects parameters in the subsequent section.

3. Best Linear Unbiased Predictors of Random Effects

Following Ma and Jørgensen,[15] we can predict the time-specific random effects $U_1, \ldots, U_t, \ldots, U_T$ by the following orthodox best linear unbiased predictors (BLUP) of U given Y:

$$\begin{aligned}
\hat{U} &= E(U) + \text{cov}(U, Y)\text{cov}^{-1}(Y)\{Y - E(Y)\} \\
&= E(U) + \text{cov}(U)B'\text{cov}^{-1}(Y)\{Y - E(Y)\},
\end{aligned} \tag{5}$$

where $\text{cov}^{-1}(Y)$ is the inverse of the covariance matrix of Y and $\text{cov}(U, Y) = \text{cov}(U)B'$, where B is a $T \times T$ diagonal matrix of $(\mu_1, \ldots, \mu_t, \ldots, \mu_T)$. In (5), $E(U)$ is a $T \times 1$ matrix which can be defined as $(1, \ldots, 1, \ldots, 1)'$ and $\text{cov}(U)$ is a $T \times T$ variance-covariance matrix of the unobserved random effects U as

$$\text{cov}(U) = \tau^2 \begin{bmatrix} 1 & \rho & \rho^2 \ldots \rho^{T-1} \\ \rho & 1 & \rho \ldots \rho^{T-2} \\ \vdots & \vdots & \vdots \ldots \vdots \\ \rho^{T-1} & \rho^{T-2} & \ldots \ldots 1 \end{bmatrix}. \tag{6}$$

The linear unbiased predictor in (5) minimizes the mean-squared distance between the random effects U and their predictors within the class of linear functions of Y.[15] In next section, we will use these linear predictors of U to construct the estimating equations for estimating the regression and random effects parameters.

4. Estimation of Parameters

4.1. *Estimation of regression parameters*

In this subsection we first consider the estimation for the regression parameters assuming that the random effects parameters are known. In next subsection, we will discuss the estimation techniques for the random effects parameters. To estimate regression parameters under known random effects parameters, we follow Ma and Jørgensen[15] and differentiate the partially observed 'joint' log-likelihood of the Tweedie mixed model for the data and random effects with respect to β yielding the partially observed 'joint' score function. Then we replace the random effects with their BLUP predictors to obtain an unbiased estimating equation for the regression parameters β, which can be expressed as

$$\psi(\beta) = \sum_{t=1}^{T} x_t' \frac{\mu_t^{1-q}(\beta)}{\epsilon^2} \left[y_t - \hat{U}_t(\beta)\mu_t(\beta) \right] = 0 \tag{7}$$

where the tth component corresponds to the response at the tth time point. Without loss of generality, we assume that the matrix of the covariates $X' = (x_1, \ldots, x_t, \ldots, x_T)$ is of full rank. Under mild regularity conditions, the solution of the estimating equation $\psi(\beta) = 0$ is consistent and asymptotically normal with asymptotic mean β and asymptotic covariance $V(\beta)$, where $V(\beta) = -S^{-1}(\beta)$ with $S(\beta) = E_\beta \{\partial \psi(\beta)/\partial \beta\}$ being the sensitivity matrix. In addition, this estimating equation $\psi(\beta) = 0$ is optimal in the sense that it attains the minimum asymptotic covariance for the estimator of β within the class of all linear functions of $\boldsymbol{Y}$.[15] This estimating equation can be solved iteratively using the following scoring algorithm,

$$\beta^* = \beta - \mathrm{S}^{-1}(\beta)\psi(\beta),$$

where the explicit expression of the sensitivity matrix is given by

$$\mathrm{S}(\beta) = \boldsymbol{X}'\mathrm{diag}\left\{\mathrm{E}(\boldsymbol{Y})\right\}\mathrm{cov}^{-1}(\boldsymbol{Y})\mathrm{diag}\left\{\mathrm{E}(\boldsymbol{Y})\right\}\boldsymbol{X}.$$

4.2. Estimation of random effects parameters

To estimate the regression parameter β, in the previous subsection we assumed the random effects parameters are known. In this subsection, we present a moment approach to estimate the unknown random effects parameters τ^2 and ϵ^2. To do that we assume that the correlation structure of the random effects are known. In next subsection we present the estimation of the correlation parameter under the AR(1) correlation structure. To estimate random effects τ^2 and ϵ^2, we use the BLUPs of the random effect $\hat{U}$. After some algebraic calculation, the iterative equations for estimating τ^2 and ϵ^2 can be expressed as

$$\hat{\tau}_r^2 = \frac{1}{T}\sum_{t=1}^{T}\left\{(\hat{U}_t - 1)^2 + \hat{\tau}_{r-1}^2 + \mathrm{cov}(U_t, \boldsymbol{Y})\mathrm{var}(\boldsymbol{Y})^{-1}\mathrm{cov}(\boldsymbol{Y}, U_t)\right\} \quad (8)$$

and

$$\hat{\epsilon}_r^2 = \frac{1}{T}\sum_{t=1}^{T}\frac{1}{\mu_t^q}\left\{(y_t - \mu_t\hat{U}_t)^2 + \mu_t^2\tau^2 - \mu_t^2\mathrm{cov}(U_t, \boldsymbol{Y})\mathrm{var}(\boldsymbol{Y})^{-1}\mathrm{cov}(\boldsymbol{Y}, U_t)\right\},$$

$$(9)$$

respectively, where $\hat{\tau}_r^2$ and $\hat{\epsilon}_r^2$ denote the estimated values of τ^2 and ϵ^2 at the rth iteration, respectively. The explicit forms for the estimators of τ^2 and ϵ^2 are similar to those presented in Ma and Jørgensen.[15] In the next subsection we present the estimation of the correlation parameter ρ.

4.3. Estimation of correlation parameter

In this section, we present a moment approach to estimate ρ under AR(1) correlation structure using the BLUP of the random effect $\hat{U}$. We can estimate the correlation parameter under AR(1) structure by using the following moment approach:

$$\rho = \frac{\text{cov}(U_t - 1, U_{t+1} - 1)}{[\{\text{var}(U_t - 1)\}\{\text{var}(U_{t+1} - 1)\}]^{1/2}}$$

$$= \frac{\text{cov}(\hat{U}_t - 1, \hat{U}_{t+1} - 1) + b(t, t+1)}{[\{\text{var}(\hat{U}_t - 1) + b(t, t)\}\{\text{var}(\hat{U}_{t+1} - 1) + b(t+1, t+1)\}]^{1/2}}, (10)$$

which can be calculated as

$$\hat{\rho} = \frac{\sum_{t=1}^{T-1}\left\{(\hat{U}_t - 1)(\hat{U}_{(t+1)} - 1) + b(t, t+1)\right\}}{\left[\left\{\sum_{t=1}^{T-1}(\hat{U}_t - 1)^2 + b(t, t)\right\}\left\{\sum_{t=1}^{T-1}(\hat{U}_{(t+1)} - 1)^2 + b(t+1, t+1)\right\}\right]^{1/2}},$$

$$(11)$$

where $b(t, t')$ is the correction term which can be simplified as

$$b(t, t') = \rho^{|t-t'|}\tau^2 - \text{cov}(\hat{U}_t, \hat{U}_{t'}).$$

The above mentioned estimation techniques for estimating regression and random effects parameters will be used in next section to analyze time series data with covariates for discrete and continuous responses.

5. Application

As mentioned earlier, our proposed random effects models can analyze time series responses with discrete and continuous margins by changing the value of index parameter q. To demonstrate this in this section we apply our approach to a count data set in daily counts of emergency room visits due to asthma from Prince George, British Columbia using $q = 1$ and a right skewed continuous data on air pollution of monitored ozone levels using $q = 2$.

5.1. Daily counts of emergency room visit data

In this section, we illustrate our proposed method with the reanalysis of the time series responses recorded as the daily counts of emergency room visits due to asthma by residents to the single hospital in Prince George, British Columbia. To check the effects of various covariates on the count responses, the data on temperature, maximum relative humidity, minimum relative humidity, TRS (total reduced sulphur) and TSP (total suspended particulates) are also collected on a daily basis. The daily measures on temperature, maximum relative humidity and minimum relative humidity were collected at the Prince George airport, which refer to the daily average readings in degrees Celsius, largest reading and smallest reading of humidity, respectively. We considered two air quality variables TRS and TSP as in Jørgensen*et al.*[10] These two air quality variables refer to the daily average readings collected from the six stations at Prince George. To consider effects of the air quality of previous days, we considered the lag 0, 1 and 2 of log TRS and log of TSP .[10] We also considered the effect of different days of the week in the initial analysis and found out that only weekends are significantly different than weekdays. In the final data analysis we incorporated an indicator variable (Day) which is 1 if the response is recorded on the weekend or 0 otherwise. Therefore the covariates considered in our model are temperature, sum of log humidities, difference of log humidities, lag 0 of log TRS, lag 1 of log TRS, lag 2 of log TRS, log of TSP and Day.

In this paper, our scientific interest is to assess the effects of the covariates on the emergency room visits due to asthma while accounting for the serial correlation and additional over-dispersion which commonly occurs in time series count data. Let Y_t represent the observed number of daily emergency room visits at the tth time point, which can be analyzed using our model with the Poisson mean parameter being specified as

$$\mu_t = \exp(\beta_0 + \beta_1 \text{ temperature} + \beta_2 \text{ sum of log humid.} + \beta_3 \text{ diff. of log humid.}$$
$$+\beta_4 \text{ lag 0 of log TRS} + \beta_5 \text{ lag 1 of log TRS} + \beta_6 \text{ lag 2 of log TRS}$$
$$+\beta_7 \text{ log of TSP} + \beta_8 \text{ Day}),$$

Table 1. Estimates of parameters with standard errors for emergency room visits data.

	Estimate	Sd. Error[a]
Intercept	-0.6224	1.7812
Temperature	0.0012	0.0058
Sum of log humid.	0.0299	0.1859
Diff. of log humid.	0.1160	0.2719
lag 0 of log TRS	-0.0221	0.0598
lag 1 of log TRS	0.0433	0.0641
lag 2 of log TRS	0.0395	0.0579
log of TSP	-0.0509	0.1211
Day	0.4594	0.0933
τ^2	0.1345	
ρ	0.3098	

[a] standard error.

Our results are presented in Table 1. Our analysis shows that temperature, sum of log humidities, difference of log humidities, lag 0, 1 and 2 of log TRS and log of TSP are appeared to be insignificant. Using their parametric model, Jørgensen et al.[10] concluded that only lag 2 of log TRS is mildly significant. We also found out that the number of emergency room visits is significantly higher on the weekends as compared to the weekdays. This is in agreement with the analysis in Jørgensen et al.[10] One of the possible reasons for higher emergency room visits during weekends is probably due to the unavailability of family doctors.

The estimates of τ^2, and ρ are 0.1345 and 0.3098, respectively. The serial correlation of 0.3098 indicates that the serial dependence of the daily counts between consecutive days is moderate. An estimate of $\tau^2 = 0.1345$ indicates that there is some variation in the responses beyond what can be characterized by the temporal random effect.

5.2. Los Angeles air pollution data

To illustrate our methodology to a rightly skewed continuous time series data set, we analyze Los Angeles air pollution data discussed in Shumway et al.[18] The raw data in their analysis included daily measurements of maximum ozone level, maximum daily temperature and daily average relative humidity in Los Angeles County during the 10 years period between 1970–1979. As the raw data had missing measurements, in their analysis of the effects of pollutants on daily mortality, Shumway et al.[18] reduced the raw data to 508 points for each of the variables which can be interpreted as smoothed weekly data.

In our analysis Y_t ($t = 1, 2, \ldots, 508$) represents the observed weekly ozone level in the Los Angeles air pollution data. Our scientific interest is to examine the effects of temperature and relative humidity on the ozone level using the reduced weekly data.[18] Our initial analysis did not find any obvious trend in the continuous time series responses. To incorporate the periodicity, we include the sine and cosine of week in the linear predictor:

$$\mu_t = \exp\left(\beta_0 + \beta_1 \text{ temperature} + \beta_2 \text{ relative humidity}\right.$$
$$\left. + \beta_3 \text{ cosine of week} + \beta_4 \text{ sine of week}\right).$$

For the index parameter q of the Tweedie family, we applied the function "tweedie.profile" in the R package [4,17] to the air pollution data. The maximum likelihood estimate of q is 2.065. We therefore take $q = 2$ which corresponds to the gamma distribution. The results of our analysis are displayed in Table 2. Our results indicate that the covariates temperature and

Table 2. Estimates of parameters with standard errors for the air pollution data.

	Estimate	Sd. Error[a]
Intercept	-2.7915	0.2098
Temperature	0.0559	0.0022
Relative Humidity	0.0107	0.0011
cosine of week	-0.0223	0.0350
sine of week	0.1835	0.0271
τ^2	0.0170	
ρ	0.8915	
ϵ^2	0.0423	

[a] standard error.

relative humidity have positive significant effects implying that the increase of temperature and relative humidity will result the increase of ozone level in the air. Our analysis also indicates that sine of the time points has significant effect whereas cosine of the time points does not have significant effect. This means the sine of the time points alone is sufficient to capture the periodicity among the time series responses.

The estimates of τ^2, ρ and ϵ^2 are 0.0170, 0.8915 and 0.0423, respectively. The serial correlation of 0.8915 indicates that the serial dependence of the time series responses is very strong. The small values of τ^2 and ϵ^2 indicate that there is not much additional variation in the responses beyond what can be characterized by the temporal random effect.

6. Discussion

In this paper, we have proposed a Tweedie generalized mixed models for discrete and continuous time series responses. Our model is more flexible than the traditional approaches as it can account various types of data. Our approach also considers the distribution-free random effects for TGLM. To incorporate the temporal dependence among the time series responses, we introduced an autoregressive of order 1 (AR(1)) correlation structure. The application of the proposed methodology on the daily emergency room visits data indicates the importance of capturing additional over dispersion among the time series responses beyond that can be captured by temporal dependence correlation structure. Our model can unify the conditional and marginal modelling approaches as the expectation of the conditional mean of the time series responses of our model is equal to the marginal mean. To estimate regression and random effects parameters, we proposed estimation techniques based on the orthodox best linear unbiased predictors (BLUP) of the random effects. Our data analysis indicates that the proposed estimation techniques perform very well.

Ackowledgements

This research was partially supported by grants from the Natural Sciences and Engineering Research Council of Canada. The authors would also like to thank Professor Yogendra P. Chaubey and the referees for their helpful comments.

References

1. K. Chan and J. Ledolter, *Journal of the American Statistical Association* **90**, 242 (1995).
2. R. Davis, W. Dunsmuir and S. Streett, *Biometrika* **90**, 777 (2003).
3. P. Diggle, K. Liang and S. Zeger, *Analysis of longitudinal data* (Oxford University Press, Oxford, 1994).
4. P. Dunn, *tweedie: Tweedie Exponential Family Models*, R Package version 2.0.7 (2010).
5. J. Durbin and S. Koopman, *Journal of the Royal Statistical Society, Series B* **62**, 3 (2000).
6. L. Fahrmeir and G. Tutz, *Multivariate Statistical Modelling Based on Generalized Linear Models* (Springer Verlag, 1994).
7. P. Heagerty and S. Zeger, *Statistical Science* **15**, 1 (2000).
8. R. Jung and R. Liesenfeld, *Allgemeines Statistisches Archiv* **85**, 387 (2001).
9. B. Jørgensen, *Journal of the Royal Statistical Society, Series B*, 127 (1987).
10. B. Jørgensen, S. Lundbye-Christensen, X. Song and L. Sun, *Statistics in Medicine* **15**, 823 (1996).
11. B. Jørgensen, *The Theory of Dispersion Models* (Chapman & Hall/CRC, 1997).
12. B. Kedem, *Time Series Analysis by Higher Order Crossings* (IEEE Press, 1994).
13. W. Li, *Biometrics* **50**, 506 (1994).
14. R. Ma, D. Krewski and R. Burnett, *Biometrika* **90**, 157 (2003).
15. R. Ma and B. Jørgensen, *Journal of the Royal Statistical Society, Series B (Statistical Methodology)* **69**, 625 (2007).
16. I. MacDonald and W. Zucchini, *Hidden Markov and Other Models for Discrete-Valued Time Series* (Chapman & Hall/CRC, 1997).
17. R Development Core Team, *R: A Language and Environment for Statistical Computing.* (R Foundation for Statistical Computing, Vienna, Austria, 2011). ISBN 3-900051-07-0.
18. R. Shumway, A. Azari and Y. Pawitan, *Environmental Research* **45**, 224 (1988).
19. S. Zeger, *Biometrika* **75**, 621 (1988).
20. S. Zeger and B. Qaqish, *Biometrics* **44**, 1019 (1988).